PALMS OF THE SOUTH-WEST PACIFIC

Howea belmoreana. (two-thirds natural size)

PALMS OF THE SOUTH-WEST PACIFIC

Their Origin, Distribution and Description

John L. Dowe

Morphology, Distribution and Ecology of Palms in New Caledonia

T. Jaffre and J.-M. Veillon
ORSTOM, Noumea, New Caledonia

Vanuatu Palms: Their Distribution and Uses

P. Cabalion
ORSTOM, Paris, France

Fiji's Native Palms and Their Potential for Landscaping

R.H. Phillips
Horticulturist, Suva, Fiji

New Caledonia Palms in Cultivation

J.L. Dowe

Illustrations by John Nelson
Translations by Marie Trefeu

Publication Fund, Palm and Cycad Societies of Australia

For Lorraine and her boys.

Palm and Cycad Societies of Australia,
P.O. Box 1134,
Milton,
Queensland 4064,
Australia

ISBN 0-9587931-3-1

Front Cover: *Clinostigma exorrhizum* on Des Voeux Peak, Taveuni, Fiji.
Back Cover: *Hedyscepe canterburyana* on Lord Howe Island (JP).

Typesetting by Art O'Graphics, Brisbane, Australia
Printing by Webb and Sons, Brisbane, Australia

CONTENTS

PREFACE

Following on with the original themes that have been a distinction of all of the Palm & Cycad Societies of Australia publications to date, Palms of the South West Pacific is our first major work that transgresses several countries.

To the uninitiated, palms in this region would conjure up no more than the scenes of coconut fringed lagoons that we are all so used to on tourist brochures. However, a look inside this work wi!! soon reveal many more species than most would realise.

Were you aware that New Zealand has an endemic (unique to the area) palm species?

Did you know that New Caledonia is second in the world (behind Madagascar – but that's another story) with the greatest number of endemic palm genera?

That the study area contains *Carpoxylon*– a genus which was only rediscovered by the author in 1987 while researching material for this book. The genus had never been properly described, had never been photographed and was thought extinct after having been lost to botanists for almost 100 years.

An absorbing work; and really the first major publication on the Palms of the South West Pacific. It not only gives detailed taxonomic accounts but also attempts to draw historical relationships and discusses the evolution of the palms of the area. Contributing authors have given accounts on local uses of the palms and the cultivation of many of them.

Thanks to John for devoting almost two years worth of spare time to the successful writing and co-ordination of the project and many thanks to the hundreds of members who have given support to our pre-publication sale. I'm sure you will agree that it was worth the wait.

Greg Cuffe
Publications Officer, P.A.C.S.O.A.
October 1989

Generous donations toward the Colour-page Scheme were received from:

Mr. and Mrs. Ian and Norma Edwards, Sydney, N.S.W.
Mr. Roger Hunt and Ms Barbara Keogh, Humpty Doo, N.T.
Mr. Colin McQueen, Broken Hill, N.S.W.
Mrs. Pauline Sullivan, Ventura, California, U.S.A.
Mr. and Mrs. Stan and Jane Walkley, Burpengary, Qld.
International Palm Society, U.S.A.
Palm Society of South Africa
North Queensland Palm Society
Palm Society of New South Wales
Palm Society of The Northern Territory
Palm and Cycad Society of Australia, Southern Qld. Group
Palm Society of Western Australia
Sunshine Coast Palm Society

AUTHOR'S PREFACE

Studies of the palms of the archipelagos and individual islands of the South-West Pacific have previously been undertaken by various researchers and authors. To date, however, a treatment of the Palm Flora (when considered as a unit defined by geographical limits) has not been undertaken, and this work provides an introduction.

There are 77 formally described palms in 36 genera in the region. They could be characterised as being occupants of moist habitats at moderate altitudes: moderately tall, predominantly arecoid with crownshafts and pinnate leaves. But in consideration of the diversity of origins and evolution of the palms, a number of other types have also developed here – the coryphoid genera *Licuala, Pritchardiopsis* and *Pritchardia,* and the calamoids with the massive monocarpic *Metroxylon* and the armed rattans of *Calamus.* Small understorey types are represented in the genera *Balaka* and *Basselinia,* the latter containing the only caespitose species in the study area. A few of the palms form extensive colonies, whilst many species are uncommon or rare components of their habitats. The affinities of the palms are varied. The subtribe Iguanurinae, in which there are 28 genera in the Old World Tropics, has reached a high level of diversity within the region. Of the 36 indigenous genera, 20 with 38 species are within that subtribe, 11 of which are restricted to New Caledonia, with seven of these being monotypic genera.

The palms of subtribe Ptychospermatinae, common or with many species in New Guinea and the Solomon Islands to the north-west, also display a marked degree of diversity. The genera of *Veitchia* and *Balaka* together total 17 indigenous species, almost one-quarter of the total species number for the region.

Apart from subtribe Archontophoenicinae, with six indigenous genera including nine species and one subspecies, the remaining subtribes of Livistoninae, Metroxylinae, Calaminae, Linospadicinae, Arecinae and Butiinae have only a limited representation.

Many of the palms are of particular interest to phytogeographers as they may indicate by association the origins, past distributions, migrational paths and evolution of the present-day flora of the region.

This work introduces and briefly discusses theories of origin and past distribution; gives present distribution at both subtribal and generic level and provides outlines to subtribes, genera and species where necessary. In addition, it includes ecological, habitat notes and detailed descriptions of the 77 valid taxa and a small number of, as yet, not formally described species. Cultivation notes are also included in the appendices.

ACKNOWLEDGEMENTS

Thanks are extended to Drs. Jean-Marie Veillon, Tanguy Jaffre and Pierre Cabalion and Mr. Dick Phillips for their contributions of their articles and papers.

Considerable assistance in preparing this work was received from Mr. Adrian Barrance, Dept. of Agriculture, Livestock and Forestry, Port Vila, Vanuatu; Mrs. Jenny Birch, Queensland, Australia; Dr. Pierre Cabalion, Strasbourg, France; Dr. John Dransfield, Kew Gardens, England; Dr. A. Esler, Auckland, New Zealand; Mr. D. Endt, Auckland, New Zealand; Dr. T. Jaffre, Noumea, New Caledonia; Mrs. A. Guymer, Norfolk Island; Mr. R.H. (Dick) Phillips, Suva, Fiji; Dr. W. Sykes, Christchurch, New Zealand; Dr. N.W. Uhl, Cornell University, New York, U.S.A.; and Dr. J.-M. Veillon, Noumea, New Caledonia.

The articles by Drs. J.-M. Veillon, T. Jaffre and P. Cabalion were originally in French and these were translated into English by Mrs. Marie Trefeu of Queensland, Australia.

For typing and proof-reading the manuscripts, may I thank Ms Julie Smith and Mr. G. Breeze, both of Brisbane, Australia. Mr. John Nelson, of Brisbane, executed the illustrations of fruit and seeds, whilst Mr. Peter Venamore, of Brisbane, prepared the maps.

For production assistance I must thank Messrs. Greg Cuffe and David Tanswell of Brisbane.

For encouragement and support I must thank the members of the Palm and Cycad Societies of Australia without whom this project would not have been possible.

INTRODUCTION

The Study Area

The South-West Pacific (Fig. 2) extends from the Torres Group (Vanuatu) in the north (Lat 13.5° South) to the Chatham Islands (New Zealand) in the south (Lat. 44° South) and from Lord Howe Island (Australia) in the west (Long. 159° East) to the Lau Group (Fiji) in the east (Long. 177° West). It is more or less a rhomboidal shaped area traversing 4,200 kms of ocean from north to south and 1,900 kms from north-east to south-west. The total surface area is approximately nine million sq kms, but actual land surface only covers 317,000 sq kms. If the two large islands of New Zealand are excluded from this total, only 49,000 sq kms of land surface area would remain. It may be useful to compare this total area of the smaller islands, where all but one of the 77 palm species are to be found, with that of Tasmania at 67,800 sq kms, Panama at 75,650 sq kms or Timor at 33,900 sq kms. The north-west is bounded by the Coral Sea, the south-west by the Tasman Sea and the north-east, east and south-east by the vastness of the Pacific Ocean.

The South-West Pacific is one of the most insular regions on earth, with more than 550 islands scattered throughout. Many of the island groups, for example the Chathams and the Kermadecs, and individual islands such as Lord Howe and Norfolk, are extremely isolated and lie many hundreds of kilometres from other land of size or influence. Australia to the west, New Guinea to the north-west and New Zealand to the south are the nearest large land-masses and it is in these areas that the origins, both past and recent, of the indigenous floras are sought.

The three main regional island groups, Vanuatu, Fiji and New Caledonia, are relatively compact clusters of islands comprising of a limited number of large principal islands accompanied by numerous smaller islands. Lord Howe and Norfolk are isolated singular islands accompanied by a few offshore islets. Raoul Island, 1,000 kms due east of Norfolk, is the most northerly of the Kermadec Islands. Chatham Islands, in the far south, are a group 600 kms to the east of South Island, New Zealand. New Zealand's North and South Islands, the largest by far in the region, have a collection of islands offshore.

The region can be divided into two distinct geographical areas. Vanuatu, Fiji and New Caledonia are confined to north of the Tropic of Capricorn, consequently being tropical in most respects. A large expanse of ocean separates the scattered islands to the south. Lord Howe, Norfolk and Raoul Islands, New Zealand and the Chathams are warm temperate to temperate, although some authors refer to Lord Howe, Norfolk and Raoul as subtropical.

GEOLOGICAL HISTORY OF THE SOUTH-WEST PACIFIC

A Simple Reconstruction:

There is disagreement, both minor and basic, among researchers regarding the geological and vegetational history of the South-West Pacific, but if data interpreting fossil records, palynology (study of fossilized pollen), basic geology, seismology etc., is examined, an assessment of the palaeobotanical history can be speculated upon (refer Smith, J.M.B. et al (1982), A History of Australasian Vegetation; Coleman, P. et al (1973), The Western Pacific).

For the purpose of this study, it is logical and illuminating to take 135 million years ago as the beginning of the relevant history of the region. At this time, in the Early Cretaceous, the breakup of Gondwana commenced. Up till this time, the crustal material we now refer to as the land-masses of South America, Africa, Madagascar, the Indian sub-continent, Antarctica, Australia, New Caledonia, New Zealand and part of Indonesia/New Guinea and possibly some other smaller terrains, made up the supercontinent of Gondwana. Gradually, rifting occurred and sections of this large land-mass commenced to drift

Figure 1.: Aerial view of Ureparapara, an island of volcanic formation in the Banks Group of northern Vanuatu.

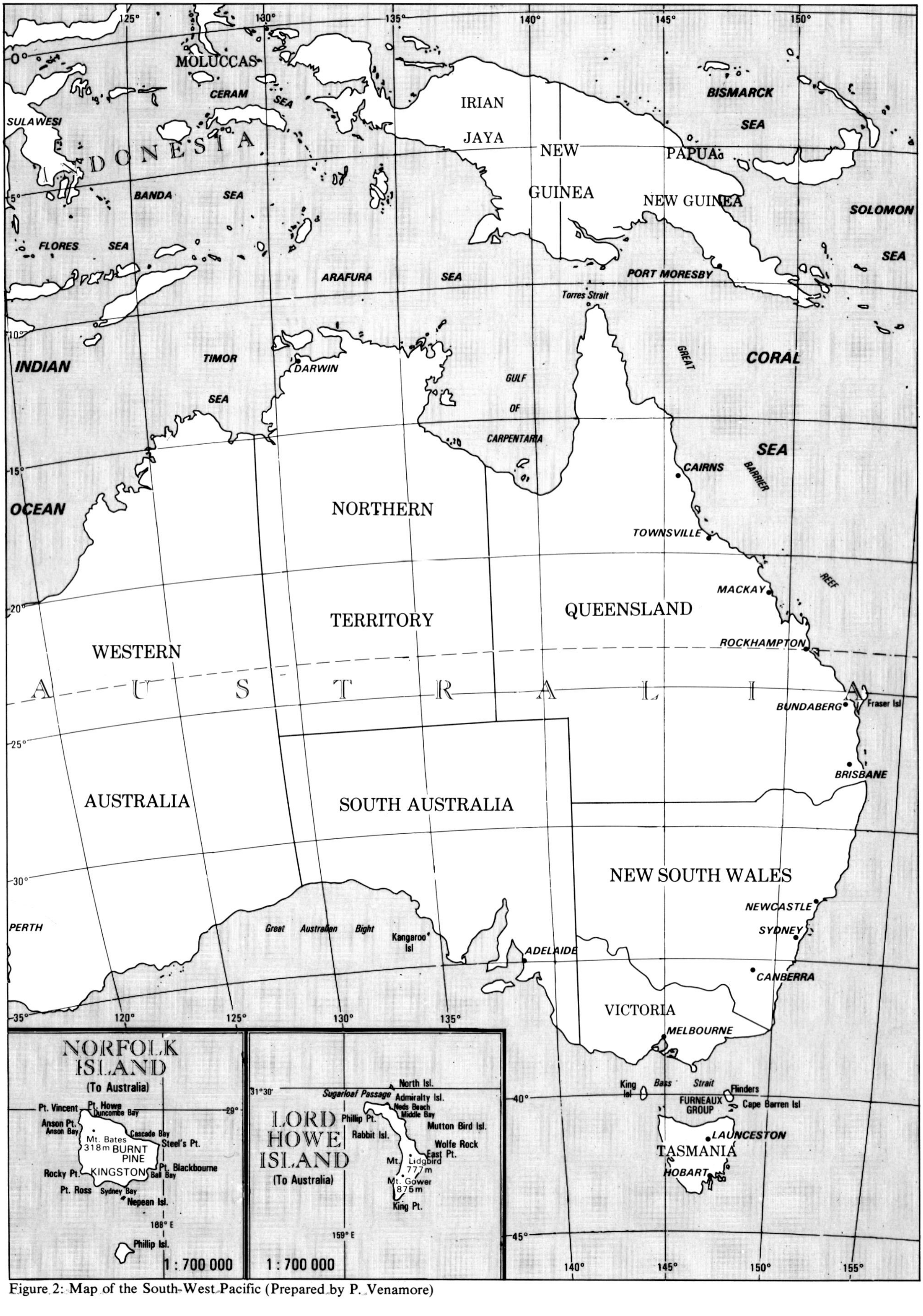

Figure 2: Map of the South-West Pacific (Prepared by P. Venamore)

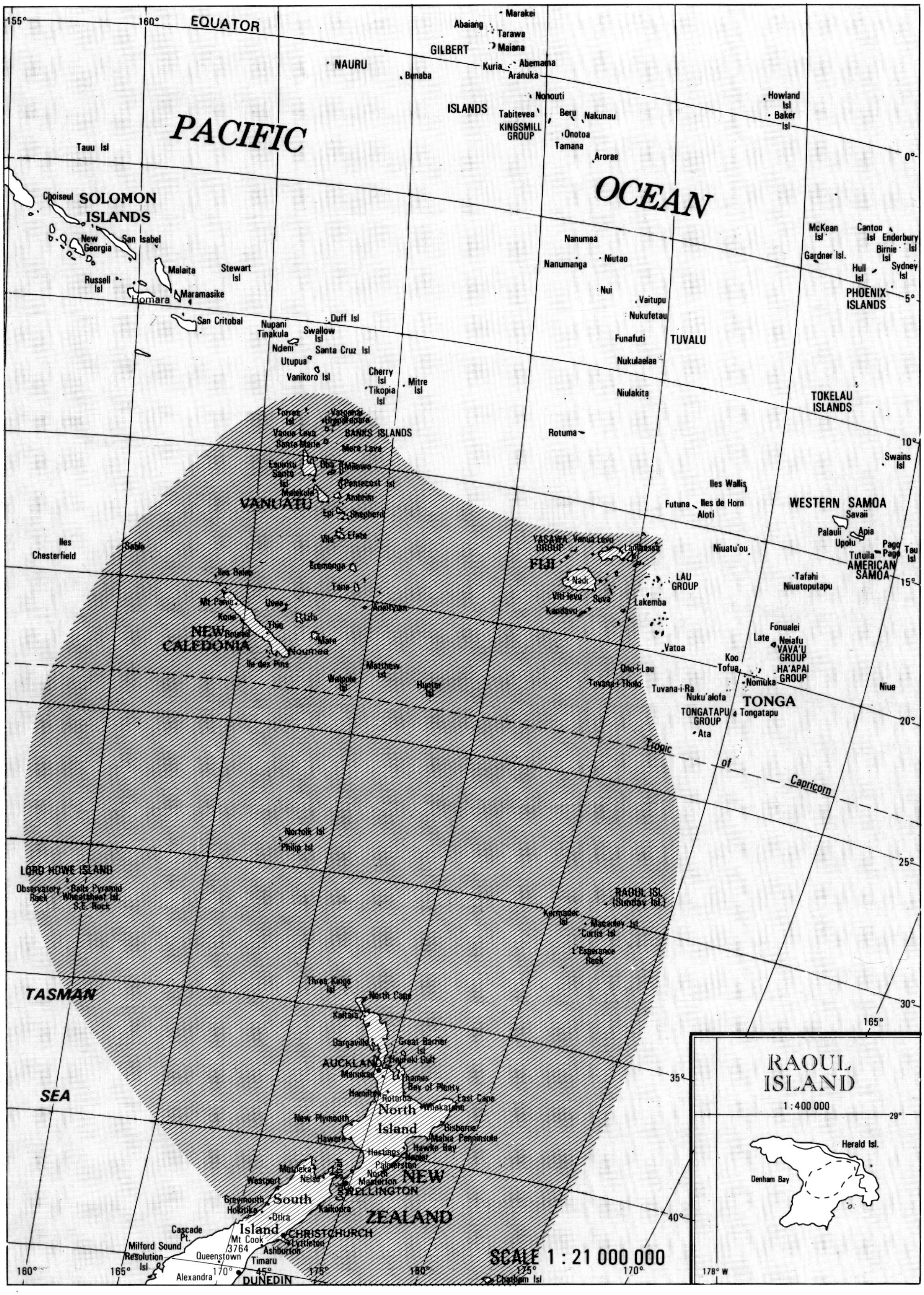

EQUATOR
PACIFIC
OCEAN
GILBERT
ISLANDS
NAURU
Banaba
Marakei
Abaiang
Tarawa
Maiana
Kuria
Abemama
Aranuka
Nonouti
Tabiteuea
Beru
Nakunau
KINGSMILL GROUP
Onotoa
Tamana
Arorae
Howland Isl
Baker Isl
Tauu Isl
SOLOMON ISLANDS
Choiseul
New Georgia
San Isabel
Malaita
Stewart Isl
Russell Isl
Honiara
Maramasike
San Cristobal
Duff Isl
Nupani
Tinakula
Swallow Isl
Ndeni
Santa Cruz Isl
Utupua
Vanikoro Isl
Cherry Isl
Tikopia Isl
Mitre Isl
Nanumea
Nanumanga
Niutao
Nui
Vaitupu
Nukufetau
Funafuti
TUVALU
Nukulaelae
Niulakita
McKean Isl
Canton Isl
Enderbury Isl
Birnie Isl
Gardner Isl.
Hull Isl
Sydney Isl
PHOENIX ISLANDS
TOKELAU ISLANDS
Rotuma
Swains Isl
BANKS ISLANDS
VANUATU
Iles Wallis
Futuna
Iles de Horn
Alofi
WESTERN SAMOA
Savaii
Apia
Palauli
Upolu
Pago Pago
Tau Isl
Tutuila
AMERICAN SAMOA
Iles Chesterfield
FIJI
Nadi
Suva
LAU GROUP
Lakemba
Niuatu'ou
Tafahi
Niuatoputapu
NEW CALEDONIA
Noumea
Vatoa
Fonualei
Late
Neiafu
VAVA'U GROUP
Koo
Tofua
HA'APAI GROUP
Nomuka
Niue
Ono-i-Lau
Tuvana-i-Ra
Nuku'alofa
TONGA
TONGATAPU GROUP
Tongatapu
Ata
Tropic of Capricorn
Norfolk Isl
Philip Isl
LORD HOWE ISLAND
RAOUL ISL (Sunday Isl.)
TASMAN
SEA
North Island
South Island
NEW ZEALAND
AUCKLAND
WELLINGTON
CHRISTCHURCH
DUNEDIN
Cascade Pt.
Milford Sound
Resolution Isl
Queenstown
Alexandra
Mt Cook 3764
Ashburton
Timaru
Chatham Isl
SCALE 1 : 21 000 000
155°
160°
165°
170°
175°
180°
0°
5°
10°
15°
20°
25°
30°
35°
40°
45°
RAOUL ISLAND
1 : 400 000
29°
Herald Isl.
Denham Bay
178° W

apart. One of the major divisions which was to separate east from west Gondwana and influence the geological and floristic history of the South-West Pacific occurred between India and Australia in about the Early Cretaceous, 100 m.y.ago. India was to eventually drift northward to collide with the northern hemisphere land-masses creating the Himalayas, thus introducing novel vegetational elements of Gondwanic origin to that region.

Meanwhile, driven by deep crustal forces, breakup of the remaining eastern land-mass into plates and platelets continued. New ocean crust was generated at some plate boundaries, whilst along the connecting boundaries, transform faulting led to relative sideways displacement of land-masses. This resulted in effective expansion to the north and north-east with some remnants becoming increasingly isolated from other parts of the original continent. On these rafts of crustal material were the ancestors of contemporary floras. No major inputs of vegetational material from other land-masses were received until the northern hemisphere was approached in the Mid Miocene (12m.y.ago). Many rafting land- masses assumed their more or less present positions in relationship to nearby areas by the Early Tertiary (50m.y.ago).

During the Mid Paleocene (60m.y.ago), volcanic activity commenced along the margins between the continental plates (upon which the remnants of Gondwana were stationed) and the Pacific plate to the north-east.

(Note: Volcanism does not occur on the junction line as it is seen at the earth's surface but from near the subducted plate boundary; thus the volcanoes are behind the leading edge of the continental plate: margin implies a zone having width behind the actual junction line.)

This collision initiated the formation of many of the islands which are still present in the region as well as some which have since been eroded, submerged or which never reached sea level at all. These are now beneath the ocean as underwater mountains, referred to as guyots, and are present on the Lord Howe and Norfolk Rises, two north-south orientated underwater ridges in the Tasman Sea.

Fiji assumed its insular form during the Early Tertiary (60m.y.ago) as a volcanic pile built upon what appears to have been the extreme north-east edge of the continental shelf which was subtending Gondwana. Fiji's exact location relative to nearby land areas at that time remains obscure, but comparative geological studies do not discount that it was within the migrational radius of the vegetation on Gondwanic land-masses to the south-west.

The Solomon Islands, the closest land-mass to the study area, were the next to be formed, assuming more or less their present position in the Early Oligocene (30m.y.ago).

At about the same time but farther south, the original volcanic formation of Lord Howe Island emerged. The northern hills of Malabar and Mount Eliza are the remnants of the remaining caldera. Further volcanic activity, adjacent to the prior emergence, in the Late Miocene (10m.y.ago), formed the southern mountains of Gower and Lidgbird.

During this interval in the Early Miocene (20m.y.ago), the older islands of Vanuatu (Espiritu Santo, Malekula and the Torres Group) emerged through the collision of the adjacent plates. Continuing activity into the Late Miocene (10m.y.ago) formed Maewo, Pentecote, parts of Epi and most of Efate.

The youngest islands in the region include Norfolk from the Mid Pliocene (2.5m.y.ago) and Raoul from the Late Pliocene (2m.y.ago) whilst during the Late Pleistocene (1m.y.ago) emerged the central and southern islands of Vanuatu which include the Banks Group, Gaua, Aoba, Paama, Ambrym, Erromango, Tanna and Aneityum and in eastern Fiji the island of Taveuni.

This simplified geological history should give at least an idea of the sequence of formation of the archipelagos and islands of the region. To discuss the complex forces which combined to bring about the above results are beyond the scope of this work. Refer to Fig.4 for a diagrammatic time scale.

Figure 3: Upper. La Roche Percee (The Pierced Rock), an eroded sedimentary formation on the west coast of New Caledonia near Bourail (RF).
Lower. The view from Little Island west to North Head and Mount Eliza on Lord Howe Island. *Howea forsteriana* can be seen on the foreshore (JP).

GEOLOGICAL CONDITIONS TODAY

The present geological condition may be summarised as follows. Arcs of semi-active volcanism delineate the boundaries of the converging continental plates. New Caledonia, Fiji and the islands south of the Tropic of Capricorn are on the Indo-Australian Plate of Gondwanic origin, whilst Vanuatu is on the Pacific Plate. The line of convergence, or subduction in some areas, extends from through the Solomon Islands in the north, southward through Vanuatu, then eastward and north-eastward to the north of Fiji, farther eastward through Samoa then abruptly turning south-south-west to the Kermadecs and to the east of New Zealand. The present volcanism in Vanuatu and the Kermadecs is a direct result of the continuing convergence of the two plates. Generally, the stronger and heavier Indo-Australian Plate is pushing against and under the Pacific Plate. As a result, the eastern edge of the Indo-Australian Plate is sliding under the rising western edge of the Pacific Plate. New Caledonia is losing altitude whilst Vanuatu is gaining.

Active volcanoes are to be found in Vanuatu on the islands of Ambrym, Tanna, Lopevi, Aoba and Gaua whilst in the Kermadecs on Raoul and Curtis Islands, the former erupting as recently as 1964. Minor volcanism is also found in the Tongariro National Park in the North Island of New Zealand.

Era	Period	Division	Geological Event	M.Y.A.	Fossil Record
Cainozoic	Quaternary	Holocene		0.01	
		Pleistocene	Vanuatu islands of Tanna and Aneityum formed		
				1.8	
	Tertiary	Pliocene	Raoul Island formed		
			Norfolk Island formed	2.5	
				5.2	
		Miocene	Mounts Gower and Lidgbird of Lord Howe Island formed		
				10	
			Contact between Australia/New Guinea and Malesia commenced		
			Vanuatu islands of Maewo and Pentecote formed		***Cocos*** in New Zealand
				24	
		Oligocene			Pollen of ***Cocos zealanica*** in New Zealand
			Solomon Islands formed	30	
				37	
		Eocene			***Arecipites*** from North Maslin Sands, South Australia
			East Papuan Province formed	45	***Rhopalostylis*** in New Zealand
			Australia effects complete isolation from Antarctica	53.5	***Nypa*** from Bass Strait
		Paleocene	Fiji islands formed	60	
				65	
Mesozoic	Cretaceous		New Caledonia and New Zealand effect isolation from Gondwana	70	
					Nypa from Perth
				84	
					First palms in the fossil record
			Gondwana begins to break up	135	
				141	First angiosperms in the fossil record

Figure 4: Time-scale Chart of Geological Events and Fossil records relevant to the South-West Pacific.

CLIMATE

The vastness of the South-West Pacific invariably creates a wide range of climatic regimes with localised peculiarities and variations.

Because most of the islands are small, they have an oceanic climate without the extremes of temperature which large land-masses experience. This is particularly the case in the islands which lie within the tropics, Vanuatu, Fiji and New Caledonia, whereas the islands which occupy the higher latitudes experience overall lower temperatures but with a greater annual range.

Rainfall shows a considerable variation when compared to temperature since it varies in both quantity and yearly amounts. For the islands within the tropics, rainfall is associated with the trade winds from the south-east with maximum rainfall between November and April. Rainfall on an island, especially large islands with high mountains, can vary considerably. The low islands tend to be drier with falls between 1,800mm and 2,000mm per annum whilst the higher islands may have falls of over 6,000mm per annum on the summits and slopes of the south-eastern sectors.

The islands situated north of 20° South have heavy rainfalls associated with the monsoonal depressions which develop between October and April. During this period, cyclones and intense tropical storms may develop and move about bringing destructive winds in excess of 200 kms/hour with periods of torrential rain. Areas south of 20° South can also be affected by cyclones passing into the vicinity, bringing high winds and heavy rains, but on a reduced intensity, to areas such as the east coast of New Zealand, Lord Howe, Norfolk and Raoul Islands. These incursions occur less frequently than in the northern regions. The chart (Fig. 5) gives an outline of temperature and rainfall for major centres.

LOCATION		Jan	Feb	Mar	Apr	May	Jun	Jul	Aug	Sep	Oct	Nov	Dec	Year
Vanuatu	°C	26.4	26.8	26.1	25.2	24.0	23.2	22.4	22.7	23.2	23.9	25.0	25.9	24.6
Port Vila	mm	272	260	312	239	148	129	103	93	104	103	161	183	2107
Fiji	°C	26.8	26.9	26.8	26.1	24.8	23.9	23.1	23.2	23.7	24.4	25.3	26.2	25.1
Suva	mm	314	299	386	343	280	177	148	200	212	218	268	313	3158
New Caledonia	°C	25.7	26.3	25.4	23.8	21.3	21.0	19.8	19.9	20.6	22.1	23.7	25.0	22.9
Noumea	mm	110	116	152	116	93	95	97	72	49	46	44	76	1066
Lord Howe Is.	°C	22.5	22.8	22.2	20.5	18.5	16.8	15.8	15.8	16.7	18.1	19.6	21.2	20.9
	mm	115	113	130	158	156	187	179	139	136	128	118	166	1675
New Zealand	°C	19.2	19.6	18.4	16.4	13.8	11.8	10.8	11.3	12.6	14.8	15.9	17.7	15.2
Auckland	mm	64	94	86	114	127	135	137	142	94	107	84	84	1268
Norfolk Is.		Mean maximum 22°C							Mean minimum 16°C					19.0
Raoul Is.		Rainfall evenly distributed												1500

Figure 5: Climate Chart for the South-West Pacific. Mean averages with Temperature in Degrees Celsius and Rainfall in Millimetres.

Figure 6: Following page. Upper. Lord Howe Island seen from the nearby Admiralty Islets to the north-east (HT). Lower. Islands of the central Yasawa chain, north-west of Viti Levu, Fiji (SG).

Figure 7: Opposite following page. Upper left. Norfolk Island with the dominant pine, *Araucaria heterophylla*.
Upper right. The estuary of the Adsone' River, south Espiritu Santo, Vanuatu, with Tangoa Island in the distance.
Lower. The central mountains of Vanua Lava, Banks Group, Vanuatu, as viewed looking northward from the village of Sola.

THE PALMS

The flora of the South-West Pacific has been studied by a number of botanists and researchers during the 220 years since botanical exploration commenced with the voyages of Cook and Banks in 1769. The first botanists to work extensively with palms in the region were Brongniart and Gris in New Caledonia and Seemann and Storck in Fiji in the 1860's. Additions were contributed by Vieillard and Pancher. In the closing decades of the 1800's, Wendland and Drude were prominent. Beccari and Burret were active during the early decades of this century whilst Pichi-Sermolli in more recent decades. Most of the present taxonomic understanding is due to the work of H.E. Moore Jr. who, prior to his death in 1980, worked intensely for 25 years on the palms (Genera Palmarum, Uhl & Dransfield, 1987). Recent taxonomic work has been undertaken also in Vanuatu by Essig, Morat, Dowe and Uhl; in Fiji by Essig, Phillips and Vodonaivalu and in New Caledonia by Mackee, Morat, Veillon, Schmid and Uhl and on Raoul Island by Sykes. H.E. Moore Jr. was very active in New Caledonia, Fiji and to a lesser extent in Vanuatu.

The addition of the authors of the synonyms, original literature, herbaria deposits, etc. will not be included in this work as other works by various authors (H.E. Moore in Flora Vitiensis Nova, A.C. Smith (1979); H.E. Moore and N.W. Uhl. The Indigenous Palms of New Caledonia (1984); H.E. Moore in Gentes Herbarum, (1957); W.R. Sykes, Kermadec Islands Flora (1977), etc.) contain thorough references.

CLASSIFICATION OF THE PALMS OF THE SOUTH-WEST PACIFIC

ARECACEAE

Subfamily I. Coryphoideae
Tribe: Corypheae
Subtribe: Livistoninae

Licuala grandis
Pritchardia thurstonii
Pritchardiopsis jeanneneyi

Subfamily II. Calamoideae
Subtribe: Metroxylinae

Metroxylon vitiense
Metroxylon salomonense
Metroxylon warburgii

Tribe: Calameae
Subtribe: Calaminae

Calamus vitiensis
Calamus sp. aff. *vitiensis*

Figure 8: *Rhopalostylis baueri* var. *cheesemanii* forest near Boat Harbour at about 250m asl on the eastern side of Raoul Island (BDS, NZGS).

Subfamily IV. Arecoideae
Tribe: Areceae
Subtribe: Archontophoenicinae

Chambeyronia macrocarpa
Chambeyronia lepidota
Hedyscepe canterburyana
Rhopalostylis sapida
Rhopalostylis baueri var. *baueri*
Rhopalostylis baueri var. *cheesemanii*
Actinokentia divaricata
Actinokentia huerlimannii
Kentiopsis oliviformis
Mackeea magnifica

Subtribe: Linospadicinae

Howea belmoreana
Howea forsteriana

Subtribe: Ptychospermatinae

Veitchia montgomeryana
Veitchia macdanielsii
Veitchia winin
Veitchia simulans
Veitchia vitiensis
Veitchia arecina
Veitchia pedionoma
Veitchia petiolata
Veitchia sessilifolia
Veitchia joannis
Veitchia metiti
Veitchia spiralis

Balaka pauciflora
Balaka microcarpa
Balaka macrocarpa
Balaka longirostris
Balaka seemannii

Subtribe: Arecinae

Gulubia cylindrocarpa
Gulubia microcarpa

Subtribe: Iguanurinae

Neoveitchia storckii
Pelagodoxa henryana
Brongniartikentia vaginata
Brongniartikentia lanuginosa
Clinosperma bracteale
Lepidorrachis mooreana
Alloschmidia glabrata
Clinostigma harlandii
Clinostigma exorrhizum
Carpoxylon macrospermum
Cyphokentia macrostachya
Moratia cerifera
Basselinia velutina
Basselinia sordida
Basselinia tomentosa

Basselinia favieri
Basselinia humboldtiana
Basselinia iterata
Basselinia porphyrea
Basselinia vestita
Basselinia pancheri
Basselinia deplanchei
Basselinia gracilis
Cyphophoenix elegans
Cyphophoenix nucele
Campecarpus fulcitus
Veillonia alba
Burretiokentia vieillardii
Burretiokentia hapala
Physokentia tete
Physokentia rosea
Physokentia thurstonii
Cyphosperma balansae
Cyphosperma trichospadix
Cyphosperma tanga
Alsmithia longipes
Lavoixia macrocarpa

Tribe: Cocoeae
Subtribe: Butiinae
Cocos nucifera

Notes on the Outlines and Descriptions

The following Outlines and Species Descriptions use as their basis the classification category of Subtribe (see boxed definition below) some of which have recently been formalized in 'An Outline of a Classification of Palms' (Dransfield and Uhl, 1986) and Genera Palmarum (Uhl and Dransfield, 1987). Previously they conformed to the suprageneric affinities of Moore but had no nomenclatural status. The usefulness of the Subtribe category in a work such as this becomes apparent when the broad geographical coverage and large number of endemic genera and species is taken into account.

The Outlines of the Subtribes and Genera are not intended to be formal keys as they are not strictly dichotomous.

A SUBTRIBE is a group of genera which, because of a number of common characters, form a natural grouping. The limits which define each SUBTRIBE may appear arbitrary, but from contemporary studies based on the work of H.E. Moore Jr. (Uhl & Dransfield, 1987) it may be concluded that the SUBTRIBES as they are presently understood form logical and consistent groupings, even though minor disharmonies do occur.

A SUBTRIBE takes the root of its name from the included genus which has the earliest validly published name but is not necessarily a typical genus of the SUBTRIBE or one from which the others have evolved.

OUTLINE OF THE SUBTRIBES OF THE SOUTH-WEST PACIFIC

Leaves palmate or costapalmate, entire or segmented
Induplicately folded
Flowers bisexual, solitary, paired or clustered
Carpels 3, free basally, united by their styles
Livistoninae

Leaves pinnate or pinnately veined
Reduplicately inserted
Ovary and fruit covered in imbricate scales
Flowers in dyads unisexual or occasionally hermaphrodite
Rachilla catkinlike, bearing either diads of a hermaphroditic and a male flower or single hermaphroditic flowers
Monocarpic
Fruit large
Metroxylinae

Dioecious, pleonanthic climbing palms with leaves extending into barbed cirri
Male inflorescence with paired or solitary unisexual flowers, female inflorescence with diads of sterile male and female flowers
Fruits smallish
Calaminae

Flowers borne in triads sometimes only basally on the rachilla
Endocarp thick or thin, lacking an operculum
Flowers borne in pits in long-pedunculate interfoliar spikes, solitary or multiple in a single leaf axil
Stigmatic remains apical
Linospadicinae

Leaflet tips praemorse
Inflorescence infrafoliar, branched to 4 orders
Male flowers with sepals less than half as long as the petals
Stamens numerous
Ptychospermatinae

Leaflet tips acute
Inflorescence with widely spreading branches, sometimes pendulous, rachilla usually pendulous
Flowers spirally inserted, male flowers only a little larger than the females
Stamens 6 or more
Archontophoenicinae

Leaflet tips bilobed
Inflorescence not widely spreading, rachilla straight
Male flowers 3 times as long as females
Stamens 6
Arecinae

Leaves occasionally pinnately veined
Inflorescence either infrafoliar or interfoliar
Peduncular bracts sometimes more than one
Stamens 6 or 12
Endocarp with an operculum
Stigmatic remains apical to basal
Iguanurinae

Infloresence enclosed in a woody plicate bract
Endocarp with 3 pores at the base
Fruit massive
Butiinae

Figure 9: The emerging inflorescence of *Howea forsteriana*.

OUTLINE OF THE GENERA OF THE SOUTH-WEST PACIFIC

Leaves costapalmate
Flowers bisexual, sessile, not in sunken pits
Carpels united by their styles
Inflorescence interfoliar
Inflorescence with primary branches borne along main axis
Leaves undivided or divided to the petiole into several ribbed segments, truncate and toothed distally
Petioles toothed, more so basally
Vanuatu

Licuala

Leaf blades divided to mid-lamina along adaxial folds
Inflorescence with long peduncle branched at the apex
Petioles not toothed
Fiji

Pritchardia

Leaves palmate
Inflorescence branched at the base, bracts tubular
Fruit large, endocarp with triangular basal projection
Endosperm intruded
New Caledonia

Pritchardiopsis

Leaves pinnate or occasionally pinnately ribbed and undivided
Ovary and fruits covered in imbricate scales
Flowers hermaphrodite or unisexual
Rachillae catkin-like
Fruit very large
Monocarpic
Vanuatu and Fiji

Metroxylon

Pleonanthic climbing rattans with leaf extending into barbed cirrus
Dioecious, inflorescences with paired or solitary male flowers or sterile male and female flowers
Seed sarcotestal, lacking an endocarp
Vanuatu and Fiji

Calamus

Small to large arborescent palms
Endocarps various
Crownshaft present, occasionally lacking
Flowers always unisexual, borne in triads, usually basal or medi-positioned with male flowers distally
Inflorescence infrafoliar bearing deciduous bracts
Fruit with fibres adherent to the inoperculate endocarp
Stigmatic remains apical
Pistillode as long as the stamens
Triads borne basally, protandrous, filaments inflexed
Stamens 6-12
Lord Howe Island

Hedyscepe

Triads borne throughout rachillae, protogynous
Stamens 9-12
Norfolk and Raoul Islands, New Zealand

Rhopalostylis

Leaves few
Filaments erect, bracteoles surrounding female flowers equal
Stamens 19-50
New Caledonia

Actinokentia

Leaves moderately numbered
Pistillode shorter than the stamens, bracteoles surrounding female flowers unequal
Stamens 34-37
New Caledonia
Kentiopsis

Pistillode lacking, thickened mesocarp fibres
Stamens 19-55
New Caledonia
Chambeyronia

Pistillode shorter than the stamens, fibres adherent to the base of the endocarp only
Stamens 32-38
New Caledonia
Mackeea

Crownshaft absent
Inflorescence interfoliar, spicate
Prophyll persistent
Triads borne in pits
Endocarp free from the seed
Stamens 30-70
Lord Howe Island
Howea

Leaflets 1 ribbed, truncately or obliquely praemorse
Crownshaft present
Inflorescence infrafoliar
Peduncular bract similar to and included within the prophyll
Male flowers slightly larger than the female, both with rounded petals and sepals
Seed terete in cross section
Leaflets with more or less parallel margins
Stamens numerous 48-130
Vanuatu and Fiji
Veitchia

Seed angled in cross section
Peduncle elongate with peduncular bract inserted well above and exserted from the prophyll
Leaflets narrowly to broadly wedge-shaped
Stamens 24-50
Fiji
Balaka

Leaflets irregularly bilobed at the apex
Seed various
Male flowers at least 3 times as large as females
Stamens 6
Vanuatu and Fiji
Gulubia

Endocarp with a detaching cap over the embryo (Lacking spines or armature)
Leaf sheath splitting opposite the petiole, not forming a crownshaft
Triads extremely basal on the rachillae with paired male flowers vertically orientated distally
Fruit large
Stigmatic remains apical
Fiji
Neoveitchia

Leaves undivided
Inflorescence interfoliar
Triads basal with paired or solitary male flowers horizontally orientated distally
Fruit very large, covered with corky tubercles
Stigmatic remains basal
Vanuatu
Pelagodoxa

Leaves divided
Fruit mostly smooth, very small to large
Peduncle of inflorescence equal to or longer than the rachis
Peduncular bract exceeding the prophyll, both bracts persistent
Peduncular base remaining as a fibrous stub
Fruit ellipsoidal
Seed with basal embryo
New Caledonia

Brongniartikentia

Peduncle of inflorescence shorter than the rachis
Peduncular bract enclosed within the prophyll, both bracts caducous
Fruit globose with lateral stigmatic remains
Inflorescence infrafoliar
New Caledonia

Clinosperma

Fruit with subapical stigmatic remains
Seed with sub basal embryo
Lord Howe Island

Lepidorrachis

Crownshaft present
Fruit elongate
Fruit with apical stigmatic remains
Endocarp pitted, operculum elongate
Seed with lateral embryo
Inflorescences numerous
New Caledonia

Alloschmidia

Fruit not elongate
Endocarp not pitted
Inflorescence glabrous
Male flowers with a short trifid pistillode
Thickened fibres in the fruits
Stilt roots present
Fiji and Vanuatu

Clinostigma

Endocarp woody, fruit large
Fruit with subapical stigmatic remains
Male flowers often distal in triads
Complete peduncular bracts two
Vanuatu

Carpoxylon

Endocarp smooth, thick and vitreous
Fruit ellipsoidal with basal stigmatic remains
Male flowers with sepals quarter as long as petals
Stamens 12
New Caledonia

Cyphokentia

Stamens 6
Fruit ellipsoidal with lateral stigmatic remains
Male flowers with sepals half as long as petals
Endocarp lightly patterned
New Caledonia

Moratia

Prophyll incompletely encircling the peduncle at attachment, open abaxially
Endocarp thin and vitreous
Seed round or infrequently bilobed
Fruit small
New Caledonia

Basselinia

Fruit oblong with apical stigmatic remains
Pistillode longer than stamens
New Caledonia and Loyalty Islands
Cyphophoenix

Fruit ovoid with curved apex
Pistillode shorter than stamens
Stilt roots present
New Caledonia
Campecarpus

Endocarp and seed irregularly furrowed, ridged or sculptured
Stilt roots lacking
Fruit subglobose with subapical stigmatic remains
Epicarp minutely papillate
New Caledonia
Veillonia

Epicarp Smooth
Pistillode shorter than the stamens
Seed with a dorsal groove and 2 flanking ridges
Mesocarp not easily separated from the endocarp
New Caledonia
Burretiokentia

Seed with a dorsal ridge
Mesocarp easily separated from the endocarp
Fiji and Vanuatu
Physokentia

Leaves occasionally pinnately veined, peduncle elongate equal to or exceeding the rachis
Inflorescence interfoliar
Crownshaft absent
Prophyll and peduncular bract persistent
Fruit globose
Pistillode longer than the stamens
Rachillae lacking a bare basal portion
Vanuatu, Fiji and New Caledonia
Cyphosperma

Leaves pinnate
Prophyll completely encircling the peduncle at attachment
Petiole long
Rachillae with bare basal portion
Endocarp with thickened extended fibres
Fiji
Alsmithia

Crownshaft present
Inflorescence infrafoliar
Bracts caducous
Fruits large, with lateral stigmatic remains
New Caledonia
Lavoixia

Crownshaft absent
Peduncular bract plicate, woody and persistent
Fruits very large
Endocarp boney with 3 basal pores
Pantropical
Cocos

Incompletely known genus
Fiji
Goniocladus

Figure 10: Following page. Upper. *Licuala grandis* in lowland forest, south Espiritu Santo, Vanuatu.
Lower. Leaf profile of *Licuala* sp. 'Vanua Lava', in lowland forest on the banks of Paradise River, Vanua Lava, Banks Group, Vanuatu.

Figure 11: Opposite following page. *Metroxylon salomonense* on the foreshore of a coral terrace, Vanua Lava, Banks Group, Vanuatu.

LIVISTONINAE Saakov

Subfamily: Coryphoideae
Tribe: Corypheae

Livistona
*Pritchardiopsis**
Serenoa
Pholidocarpus
Pritchardia *
Brahea
Johannesteijsmannia
Colpothrinax
Copernicia
*Licuala**
Acoelorraphe
Washingtonia
* Indigenous genera

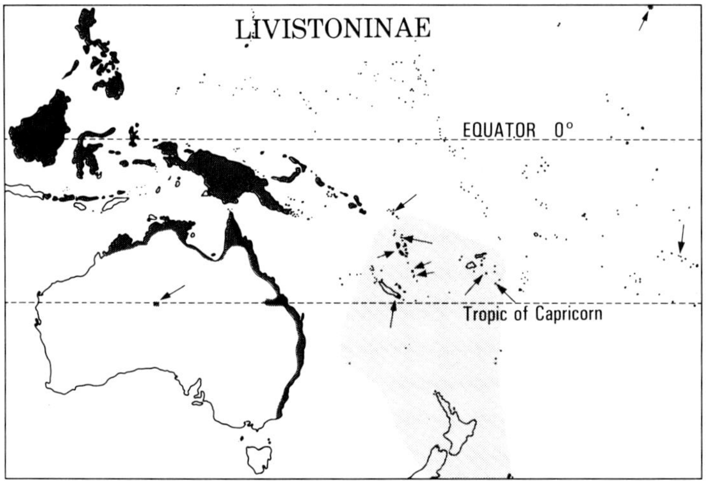

Description:
Small to large, understorey to emergent palms; solitary or clump-forming; leaf non- sheathing, palmate to costapalmate; unarmed or armed; monoecious, dioecious or hermaphroditic, pleonanthic, flowers usually bisexual, never positioned in sunken pits, fruits more or less sessile.

Distribution:
From equatorial to warm temperate; sea-level to moderate altitudes; rainforest, moist forests, coastal forests, swamp forests, stream edges, seasonally dry forests, semi-arid to arid areas, limestone to coraline islands. North-East Africa, eastern India, South-East Asia, Malesia, Philippines, New Guinea, Australia, Solomon Islands, Vanuatu, New Caledonia, Fiji, mid-Pacific island groups, Hawaii, warm temperate to tropical North America, Central America, West Indies, tropical to equatorial to warm temperate South America.

Notes:
This SUBTRIBE is found on all continents except Europe. It is poorly represented in the South-West Pacific with only *Licuala* being represented by more than one taxon. It is basically a SUBTRIBE of unspecialised palms of ancient origins with some genera, e.g. *Livistona,* and *Serenoa,* found lingering in areas where rainforest has once predominated but has either retreated or vanished for complex reasons. The small genera of *Johannesteijsmannia* and *Pholidocarpus* are restricted to equatorial rainforest whilst *Licuala* is one of the most diverse genera in ARECACEAE with at least 110 species, whilst *Pritchardiopsis,* and *Serenoa* are monotypic and of a limited distribution in tropical and subtropical regions. Only *Pritchardiopsis* is endemic to the South-West Pacific.

LICUALA Thunb. 1782

A genus of about 110 species distributed in the Old World tropics from India to the Philippine Islands, New Guinea, Australia, Solomon Islands and Vanuatu. One species in the South-West Pacific.

Licuala grandis H. Wendl. ex Linden 1881

syn. *Pritchardia grandis*

Locale, Habitat and Ecology:
Vanuatu on the islands of Efate, Malekula, Pentecote, Espiritu Santo and the Banks Group as well as the Santa Cruz Group in the Solomon Islands, where it occurs abundantly as an understorey component in primary, secondary and disturbed rainforest and open coastal forests at low to moderate elevations on moist but well drained alluvial soils, usually withdrawn from the coral terraces. In some areas, (e.g. south Espiritu Santo) this species grows in dense colonies in positions exposed to direct sun-

light between large trees, commonly *Ficus* spp. The fruits, produced in abundance, are consumed and distributed by fruit bats and large birds. The leaves are used as a food wrapping by the indigenous people. It is a very common horticultural subject.

Description:

A medium palm to about 4m with a slender, roughly scarred trunk with persistent leafbases in the upper portion. Leaves number about 12 with the petiole, to about 1 m long, firstly held erect but becoming pendulous. Leafbases do not form a crownshaft. Leaf blades are costapalmate, usually undivided, nearly orbicular in outline to 1.1m wide and 80cm long, mid to dark green above, lighter beneath, apically toothed with induplicate folds.

The inflorescences are interfoliar, extending beyond the leaves and usually pendulous, with about 6 once-branched branches. The bracts subtending the main axis and branches are tubular and persistent. Flowers are usually bisexual, greenish-brown to 5mm long, loosely grouped along the rachilla. Stamens 6, with the stamen filaments united basally in a lobed ring, 3 positioned on the crests and 3 in the sinuses. The spherical fruits are orange/ crimson at maturity, 1 to 1.5cm in diameter. The brittle endocarp is longitudinally ribbed. The homogeneous endosperm is intruded by the testa below the raphe.

Notes:

An unstudied *Licuala* occurs on Vanua Lava in the Banks Group of Vanuatu. A resemblance and possible affinity with *L.grandis* is obvious. The differences include leaves which are symmetrically divided, the central segment in which the costa is inserted is v-shaped and wide to about 30cm, the lateral segments are all similar to about 6cm wide. All segments are divided to the petiole, which is toothed only basally. The inflorescence is attributable to *L.grandis,* whereas the fruits are smaller with a thin dryish mesocarp, whilst the endocarp is sparingly ribbed. The fruit is yellowish at maturity. An interesting growth phenomenum is that the leaves commence to divide at about the 3rd or 4th juvenile leaf. Whether the differences warrant a new species description is uncertain. This taxon may only represent a form of *L.grandis* with divided leaves.

PRITCHARDIA Seem. & H. Wendl. ex H. Wendl. 1862

A genus of about 37 species distributed through the central, southern and south- western Pacific on the Hawaiian Islands, which has the majority of species, the Danger Islands, Tuamotu Archipelago, Cook Islands, Tonga, Solomon Islands and Fiji. One endemic species in the South-West Pacific.

Pritchardia thurstonii F.v.Muell. & Drude 1887

syn. *Washingtonia thurstonii*
Eupritchardia thurstonii
Styloma thurstonii

Locale, Habitat and Ecology:

Fiji, in the Lau Group, where it is endemic to the limestone/volcanic islands of Vanuablava, Ogea, Fulaga and the islets in the Fulaga Lagoon. The limestone islands and islets support a sparse flora. Erosion is severe both from rainfall and wave action, the latter dramatically undercutting many of the smaller islets. The environment on these islands is hot and dry, with strong winds and salt laden air. This species forms moderately dense, extensive colonies, and dominates the associated flora. The leaves are used to make fans. It is a well known horticultural subject.

Description:

An occasionally tall palm to 15m in height but usually less, with a grey longitudinally fissured, closely scarred trunk to 20cm in diameter. Leaves number about 20, are costapalmate with an arching, unarmed tomentose petiole to about 1m long. The leaf blade is mid-green, lightly glaucous above, lighter green beneath, to 1m long and wide, with segments divided to about 25cm, numerous and taper-

Figure 12: Following page. Upper left. *Metroxylon vitiense* in lowland swamp habitat, Ngaloa, south Vitu Levu, Fiji. Upper right. Yellow petiole with flexible combed-spines are characteristic of *Metroxylon vitiense.* Lower. Fruit of *Metroxylon salomonense* displayed upon a mat of pandanus leaf.

Figure 13: Opposite following page. Upper left. Fruiting group of *Metroxylon warburgii,* north-east Efate, Vanuatu. Upper right. Spined petiole of *Metroxylon warburgii,* on a plant from Vanua lava, Vanuatu. Lower. Fruit of *Metroxylon warburgii,* from left, ungerminated, germinated, and with epicarp removed.

ing, induplicately folded with a strong costa to 25cm long. A thick fibre is adherent to the leafbase.

The interfoliar inflorescences are pendulous, one per leaf axil, to 2m long, exceeding the leaves. Prophyll and peduncular bracts, usually 4, are tubular, becoming fibrous toward the apex. The peduncle is elongate, branched at the apex. Rachillae are to about 12cm long. Flowers are bisexual, sessile, borne singly and spirally on the rachilla, pale yellow to 7mm long. Sepals are united to form a tube. Petals are united basally, valvate above and deciduous at anthesis. Stamens 6. The carpels are basally distinct but united above by their styles. The fruit is globose, 7mm in diameter, deep red at maturity, and apically bearing the remains of the carpels and styles. The seed is globose with a basal embryo.

PRITCHARDIOPSIS Becc. 1910

A monotypic genus which is closely related to *Livistona* and *Licuala* and is endemic to New Caledonia.

Pritchardiopsis jeanneneyi Becc. 1910
syn. *Licuala jeanneneyi*
L.jeannencyi Hort.

Locale, Habitat and Ecology:

Endemic to the extreme south of La Grande-Terre, New Caledonia near the Bay of Prony at about 200m alt. in forest on ultrabasic rocks. Considered extinct due to the consumption of its 'cabbage' by convicts incarcerated nearby at the Prony Penal Settlement during the last century, this palm was rediscovered in the late 1970s. In 1980, botanists working at ORSTOM in Noumea were presented with a sample of leaf and inflorescence which were confirmed to be *Pritchardiopsis.* The only known population over about 1ha has about 30 individuals; one mature seed producing palm, the other juveniles of disparate age to the older palm.

Description:

A tallish palm to 12m with a closely ringed grey trunk to 15cm in diameter. A few leafbases with abundant matted fibre are retained below the leaves. Leaves number about 15, are palmate with a prominent hastula and small costa about 12cm long. The blade, mid green above and lighter below, is about 1m wide and long. Segments are induplicately folded, divided to mid lamina and bifid at the apex. The erect to arching petiole is unarmed, about 1.5m long and 3.5cm wide.

The glabrous interfoliar inflorescence is about 1.4m long, thrice branched, semi-pendulous. The prophyll is attached about 12cm above the base. Each branch is subtended by a rachis bract and farther branches by papery tubular bracts. Ultimate rachillae, about 8cm long, are without bracts. Flowers are bisexual, sessile, arranged in groups of 3, at least basally. Petals are fleshy, united basally then valvate, stamens number 6. The ovary is trilocular and triovulate. Petals are spreading at anthesisand persistent. Fruit is globose, to 3.5cm in diameter, black at maturity with apical stigmatic remains. Mesocarp is fleshy, the endocarp is woody, papillate, with broad longitudinal ridges on one side, rounded on the other and attenuated basally.The seed is globose with a basal hilum and orbicular raphe. The testa intrudes into the homogeneous endosperm.

Figure 14: *Pritchardiopsis jeanneneyi,* a rare and endangered species endemic to the hills above the Bay of Prony, New Caledonia (O).

METROXYLINAE
Blume

Subfamily: Calamoideae
Tribe: Calameae

*Metroxylon**
Korthalsia
* Indigenous genus

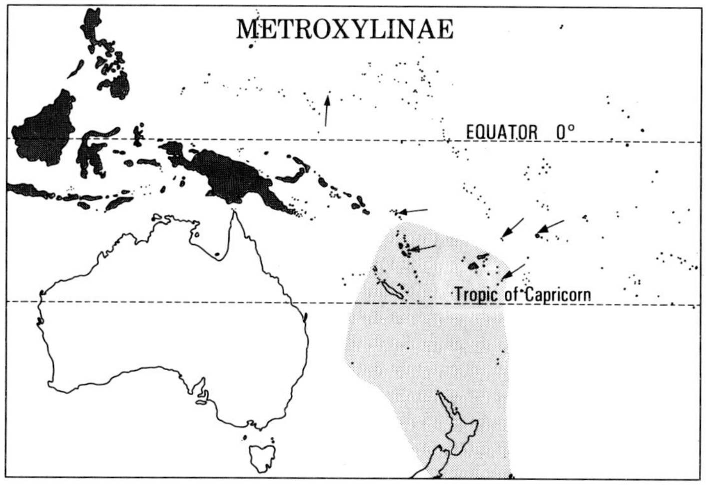

Description:
Small to massive, emergent palms; solitary, clump-forming, branching or climbing palms; leaf non-sheathing or sheathing, pinnately leaved; unarmed or armed; monoecious, hapaxanthic or monocarpic; inflorescense terminal; rachilla catkin like, fruits imbricately scaled.

Distribution:
From equatorial to tropical; sea-level to moderate altitudes; rainforest, coastal forests, swamp forests, moist forests, foreshores on coraline substratum. South-East Asia, Malesia, Philippines, New Guinea, Solomon, Guam and Caroline Islands, Vanuatu and Fiji.

Notes:
Korthalsia is a group of armed climbing palms which has the ability to form aerial branches and has an inflorescence morphology which is remarkably similar to that of *Metroxylon* and unlike any other rattan, which suggests that rattans may have evolved along two distinct ancestral lines (Dransfield 1979). *Metroxylon* is one of the few genera which has diversified to a greater degree in a region away from the considered centre of origin, New Guinea, being represented by three or four species in the western Pacific and only one or two in New Guinea.

METROXYLON Rottb. 1783

A genus of five species distributed from East Malesia, New Guinea, Philippines, Guam, Caroline Islands, Solomon Islands, Santa Cruz Group, Vanuatu, Fiji and Western Samoa. Three species in the South-West Pacific.

Key to Species

Inflorescence branched to 2 orders
    Ultimate rachillae pendulous
        Trunk to 15m in height
            Fruit turbinate to globose-ovoid
                Fiji — *M. vitiense*

Inflorescence branched to 2 or 3 orders
        Trunk to 20m in height
            Fruit globose
                Vanuatu — *M. salomonense*

Inflorescence branched to 3 orders
    Ultimate rachillae erect
        Trunk to 7m in height
            Fruit pyriform
                Vanuatu — *M. warburgii*

Figure 15: Petiole base of *Metroxylon warburgii*, with spineless pulvini.

Metroxylon vitiense (H. Wendl.) H. Wendl. ex Hook. 1882
syn. *Sagus vitiensis*
Coelococcus vitiensis

Locale, Habitat and Ecology:
Fiji, on the islands of Viti Levu, Ovalau, Vanua Levu and Rotuma, occurring commonly (at least in south Viti Levu) as an emergent in pandanus swamps *(P. pyriformis)* and nearby drier slopes. The pith of the trunk is not used as a source of sago as it is in New Guinea and Malesia but a hardy thatch is produced from the large dried leaves. Roofs and walls made from this source are reported to last up to 30 years. The palm heart, the upper section of the trunk which contains the developing leaves is infrequently consumed as a salad vegetable. It is commonly cultivated as a village plant for thatch but only rarely as an ornamental.

Description:
A large palm with a solitary brownish trunk to 15m in height and 1m in diameter. Shredded persistent leafbases protect 3cm long rootlets which protrude from the scar tissue but become exposed on the eventual shedding of the remaining leafbase fibres. The erect leaves are numerous, to 5m long, with brown/chestnut spines in oblique crests and pulvini on the non-sheathing leafbase, petiole and rachis. Dead leaves persist for a considerable time, forming a sparse untidy 'skirt'. Leaflets are numerous, to 7cm wide, acute, with a prominent mid-rib.

A monocarpic species with a terminal inflorescence to 4m high, the primary axis emerging from within the uppermost bract-like leaves. First order branches are arranged distichously, each subtended by a persistent bract, the second order branches are pendulous, catkinlike rachillae to 10cm long. Flowers, either male, female or hermaphroditic are spirally arranged in pairs with united bracteoles. Petals are valvate, male and hermaphroditic flowers have 6 stamens. The ovary is trilocular and triovulate, covered in imbricate scales. Fruits, similarly scaled, are turbinate, globose to globose-ovoid, yellowish brown, 7cm high be 5cm wide. The seed coat is invaginated.

Metroxylon salomonense (Warb.) Beccari 1913
syn. *Coelococcus salomonensis*
Metroxylon bougainvillense

Locale, Habitat and Ecology:
In Vanuatu on the island of Vanua Lava in the Banks Group, but widespread through the Santa Cruz Group, Solomon Islands and eastern New Guinea. On Vanua Lava, a small population occurs on the well drained south-eastern coral terrace fringed with mangroves *(Rhizophora sp.)*. The palms are emergent, receiving strong winds at times. *Barringtonia* and *Cyathea* spp occupy the same habitat. The leaves are harvested for thatch, mainly as roofing, by nearby villagers. Sago is not obtained from the pith nor is the palm heart consumed as is the case with *Metroxylon* ssp in New Guinea and east Malesia. Other forms of carbohydrate and salad vegetables are easily obtained and cultivated by the villagers on this fertile volcanic island.

Description:
A very large palm with a grey-brown solitary trunk to 20m in height and 1.2m in diameter. Persistent leafbases protect rootlets which emerge from the scar tissue becoming exposed on the eventual shedding of the remaining leafbase fibres. The leaves, to 11m long, are at first erect but become horizontal and then arching, the dead leaves persisting as an untidy 'skirt'. Flexible spines to 40cm long are positioned in oblique crests and pulvini on the leafbase, petiole and rachis. The petiole is massive, to 40cm in width. Leaflets are numerous, to 10cm wide, marginally undulate, acute with a prominent mid-rib.

A monocarpic species with a terminal inflorescence to 4m high, the primary axis emerging from within the uppermost bractlike leaves. First order branches are arranged spirally, horizontally held. Second order branches are pendulous, catkinlike rachillae to 20cm long. Flowers, either male, female or hermaphroditic are spirally arranged in pairs with silver-grey pubescent united bracteoles. Petals are

Figure 16: Upper left. Inflorescence branch and fruit of *Metroxylon warburgii* being held by Chanel Sam, Research Officer, ORSTOM, Port Vila, Vanuatu.
Upper right. Palm hearts extracted from *Metroxylon vitiense,* south Viti Levu, Fiji.
Lower. *Metroxylon vitiense* sharing its swamp habitat with *Pandanus pyriformis.*

valvate, male and hermaphroditic flowers have 6 stamens. The ovary is trilocular, triovulate and covered in imbricate scales. Fruits, similarly scaled, are globose, apically and basally depressed, green/straw coloured, 9cm wide and 6.5cm high. The seed coat is invaginated.

Metroxylon warburgii (Heim) Beccari 1918
syn. *Coelococcus warburgii*

Locale, Habitat and Ecology:
In Vanuatu, on the islands of Vanua Lava, Espiritu Santo, Malekula, Pentecote, Erromango and Efate as well as in Western Samoa and the Santa Cruz Group, in colonies in swamps, bordering watercourses, seepage areas to well drained slopes and coral terraces, usually as a semi-emergent but occasionally as an emergent component. As are the other *Metroxylon* spp in the South-West Pacific, the pith is not extracted as a source of raw carbohydrate, but the leaves are harvested to produce a durable thatch. This species is commonly cultivated for this purpose. The abundant pulp in the large fruit is unpalatable but the hard endosperm is occasionally used as vegetable ivory.

Description:
A medium palm with a grey/brown solitary trunk to 7m in height and 30cm in diameter, the nodes elongating from about mid trunk to the narrowing apex. The deciduous leaves, to 3m long, are held erect. Flexible, papery spines are inserted singly or in oblique crests and pulvini in the mid-green leafbase, petiole and rachis. Spineless pulvini pattern obliquely the almost encircling leafbases. Leaflets are numerous, to 6cm wide, pendulous toward the acute apex and with a prominent mid-rib.

A monocarpic species with a terminal inflorescence to 3m high, branching to 3 orders, all branches being erect with the upper side glabrous. Ultimate rachillae, to 12cm long, are catkinlike. Flowers, either male, female or hermaphroditic are large, spirally arranged in pairs. Petals are valvate, male and hermaphrodite flowers have 6 stamens. The ovary is trilocular, triovulate and covered in imbricate scales. Fruits to 12cm long and 9cm wide are similarly scaled, pyriform with a depressed base and persistent calyx. The seed coat is invaginated and has a basal embryo.

Notes:
Metroxylon within the South-West Pacific is more variable than the present taxonomy indicates. An unstudied species occurs on Erromango. It has a massive trunk, large leaves but small ovoid fruits. *Metroxylon salomonense* on Vanua Lava in the Banks Group has an inflorescence branched to two orders whereas the current literature cites it as being branched to three orders. Its leaflets also have distinct undulate margins. The height of *M. warburgii* differs from island to island, those on Efate being less than 5m whilst those on Espiritu Santo usually exceeding 7m at maturity. Of course these differences may only be ecological.

Figure 17: Leaf of *Metroxylon warburgii.*

CALAMINAE Meisner

Subfamily: Calamoideae
Tribe: Calameae

Eleiodoxa
Calospatha
Salacca
Pogonotium
Daemonorops
Ceratolobus
*Calamus**
Retispatha
* Indigenous genus

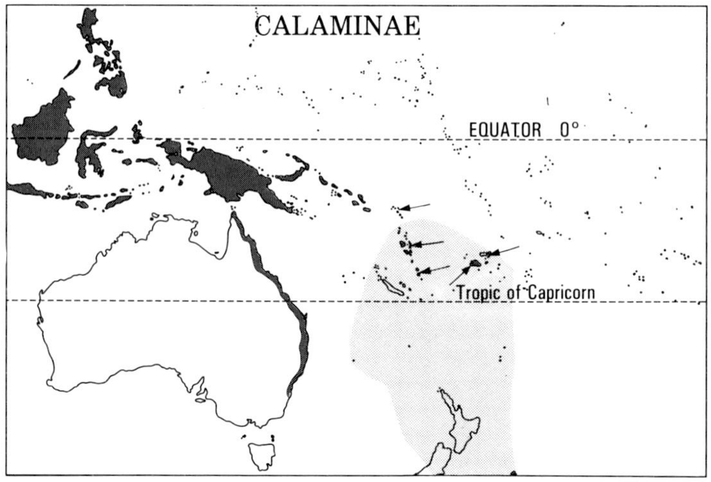

Description:
Small to large, understorey to emergent; solitary, clump-forming, acaulescent, climbing palms; leaf sheathing, pinnately leaved; armed; dioecious, pleonanthic, hapaxanthic; scaly-fruited.

Distribution:
From equatorial, tropical to warm temperate; sea-level to sub-alpine altitudes; rainforest, moist forests, coastal forests, swamp forests, stream edges, dry ridges. Africa, Nepal, north and east India, Sri Lanka, southern China, South-East Asia, Malesia, Philippines, New Guinea, north-eastern and eastern Australia, Solomons, Fiji and Vanuatu.

Notes:
Of the eight genera included in this SUBTRIBE, only *Calamus* extends into the South-West Pacific. Most of the other genera have restricted distributions in Malesia and New Guinea. *Calamus* is the most diverse genus in ARECACEAE with at least 370 species. The representation in the South-West Pacific is peripheral with only two or three species present.

CALAMUS L. 1753

A genus of about 370 species distributed from tropical Africa, India, Nepal, Sikkim, China,Sri Lanka, South-East Asia, Taiwan, Philippine Islands, Malesia, New Guinea, eastern Australia, Solomon Islands, Vanuatu and Fiji. Possibly three species in the South-West Pacific, one of which is described, another distinct and others of uncertain status.

Calamus vitiensis Warb. ex Becc. 1908

Locale, Habitat and Ecology:
Fiji, on the islands of Viti Levu, Taveuni and nearby islets and possible in Vanuatu where it is widespread, at elevations from near sea-level to 800m in moist to very moist forests as a climbing emergent palm. On Viti Levu it grows in association with such palms as *Balaka* and *Veitchia* spp and *Neoveitchia storckii.* On Taveuni it grows at elevations of 0-300m, overlapping with *Veitchia simulans* in the upper range and is associated with *Balaka seemannii* on the very wet south-east slopes. It appears to be found only on fertile volcanic soils as no collections or observations have been made from areas overlying the coral terraces or limestone.

Description:
A solitary, climbing palm with a stem to 15m or more, a diameter of 6cm with elongate internodes, particularly in juveniles. The lower portion is bare of leaves, deep-green with prominent leaf scars. A tight

mound of thin yellowish roots are exposed at the base, some aerial. Leaves, retained only in the apical few metres of the trunk, are to about 2m long, terminating in a hooked cirrus to about 1.5m long. The leafsheath is elongate, dark green with brownish indument, and dark spines arranged in short horizontal combs, or singly, denser toward the petiole. The knee is prominent, incurved at the base, spineless, absent on juvenile or immature specimens. A short ridge to 3cm long and pointed distally, is present on the upper petiole at the base. A petiole is absent or very short on mature palms, spined on the edges and above and into the rachis for a short distance. The rachis is angled, unarmed for most of its length. Evenly arranged recurved barbs emerge below near the apex and continue into the cirrus terminating in single or paired hooks. The leaflets, about 20 per side regularly arranged, are broadly elliptical, acute, deep green with two prominent veins on each side of the mid-rib measuring to 30cm long and 6cm wide, becoming pendulous toward the apex.

This is a dioecious species. The inflorescences arise from within the non- flagellate leafsheath, firstly erect, becoming pendulous in fruit, at least the female, about 2m long, the males shorter. There are about 4 branches, each with a persistent green tubular bract, remaining green in fruit. The recurved rachillae, to 12cm long, are distichously and alternately arranged. Male flowers are solitary with 6 stamens and a small pistillode. Each female is paired with a sterile flower, the ovary is trilocular and triovulate covered with imbricate scales. Fruits are similarly scaled, sessile, with apical stigmatic remains, to about 1.3cm long by 1cm in diameter, white at maturity. Seeds are basally depressed.

Notes:
The undetermined *Calamus* sp. from Vanuatu I have listed as *Calamus* sp.aff. *vitiensis.* Differences when compared to *C. vitiensis* include crustose scales on the leaf margins and mid-rib below, a longer inflorescence with more branches, spines never united basally into combs and fruits which are cream, not white on maturity. Such differences commonly occur within the variational limits of many palm species.

Calamus.. sp. aff. *vitiensis.* Fruit (natural size) and seed in two views (twice natural size)

Figure 18: Following page. Upper. *Calamus vitiensis,* cultivated in the garden of R.H. (Dick) Phillips, Suva, Fiji. Lower left. Spined leafsheath and old infructescence of *Calamus vitiensis,* on a plant growing on Taveuni, Fiji. Lower right. Maturing fruit of *Calamus vitiensis,*

Figure 19: Opposite following page. Upper. *Calamus* sp. aff. *vitiensis* is common throughout the forests of Vanuatu. Lower left, Leafsheath and petiole of *Calamus* sp. aff. *vitiensis.* Lower right. Mature infructescence of *Calamus* sp. aff. *vitiensis.*

ARCHONTOPHOENICINAE

J. Dransfield & N.W. Uhl

Subfamily: Arecoideae
Tribe: Areceae

Archontophoenix
*Kentiopsis**
*Chambeyronia**
*Mackeea**
*Hedyscepe**
*Actinokentia**
*Rhopalostylis**
* Indigenous genera

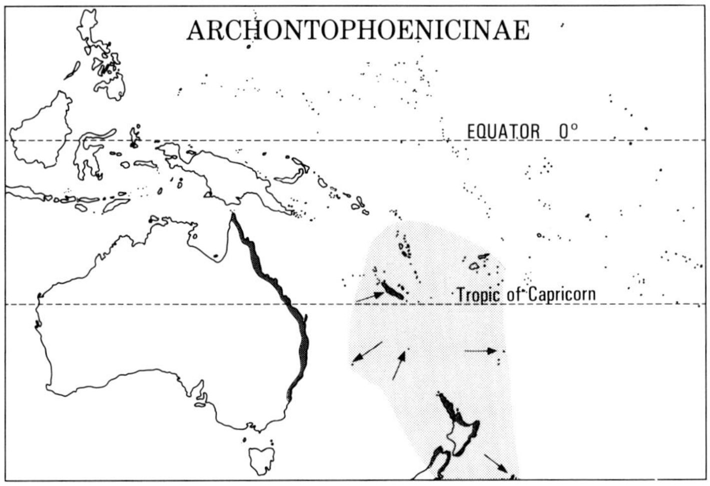

Description:
Medium to tall, usually emergent; solitary; sheathing, pinnately leaved palms; unarmed; monoecious, pleonanthic; flowers unisexual; endocarp lacking an operculum.

Distribution:
From subequatorial to temperate; sea-level to moderate altitudes; rainforest, moist forests, swamp forests, coastal forests. Eastern Australia, New Caledonia, Lord Howe Island, Norfolk Island, Raoul Island, New Zealand and Chatham Islands.

Notes:
This SUBTRIBE may be considered to be centred in the South-West Pacific as six of the seven genera are confined to the region. The genus which does not occur in the area, *Archontophoenix*, is found in subequatorial to warm temperate eastern Australia. *Chambeyronia, Kentiopsis, Mackeea* and *Actinokentia* differ from the other genera in basic leaf morphology, which suggests that they may have originated from the same evolutionary source (Uhl, 1984). This SUBTRIBE is confined to the remnants of what is considered to have been the ancient continent of Gondwana.

CHAMBEYRONIA Vieill. 1873

A genus of two species endemic to New Caledonia, closely related to *Archontophoenix* from eastern Australia and *Hedyscepe* from Lord Howe Island. *Chambeyronia macrocarpa* is the type species.

Key to Species

Leaflets broad
  Rachillae glabrous
    Male sepals much smaller than petals
    Stamens 40-55
      C. macrocarpa

Leaflets narrow
  Rachillae lepidote
    Male sepals about same length as petals
    Stamens 30-40
      C. lepidota

Chambeyronia lepidota H.E. Moore 1978
  syn. *Lavoixia lepidota*

Locale, Habitat and Ecology:
New Caledonia, being restricted to the north-east sector of La Grande-Terre at elevations of 300 to

1400m as an emergent palm in moist forests on schistose soils. This species' main population is centred on the Mt. Panie massif.

Description:
A medium palm with a closely ringed, brown trunk to 10m in height and 14cm in diameter. The arching and spreading leaves number about 10, with a slightly bulging, elongate crownshaft to 1m long, whitish green covered in red/brown scales. Regularly arranged leaflets number about 34 per side, are stiff with ramenta present on the mid rib beneath. Leaflets measure from 1cm to 6cm wide and 20cm to 1.12m long. Basal leaflets are reduced to lorae.

The inflorescences are stiffly held, densely covered with brown scales. The peduncular bract is enclosed within the prophyll which is 60cm in length and 14cm wide, both covered in lepidote scales. The angled flexuous rachillae measure to 40cm in length. Triads are borne throughout, the male flowers to 1cm high, sepals and petals of similar size, 30-40 stamens, the females to 1.1cm high with lepidote sepals and petals. The red obovoid to ellipsoidal fruits measure about 3cm long by 2cm in diameter. The endocarp lacks an operculum and the seed is globose to obovoid.

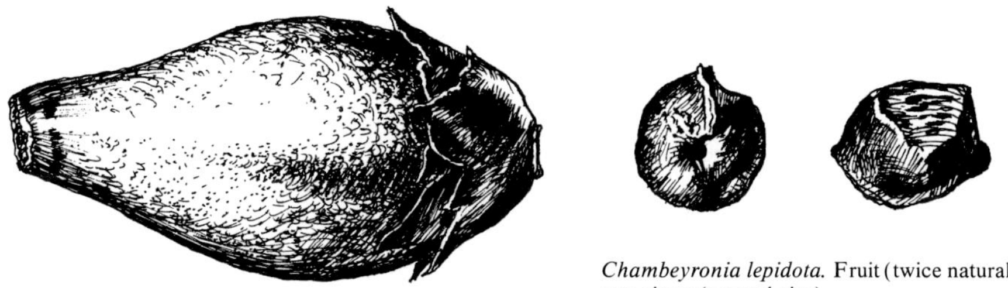

Chambeyronia lepidota. Fruit (twice natural size) and endocarp in two views (natural size)

Chambeyronia macrocarpa (Brongn.) Vieill. ex Becc. 1920
syn. *Kentiopsis macrocarpa*
Kentia lindenii
Kentia luciani
Kentia macrocarpa
Kentia rubricaulis
Chambeyronia hookeri

Locale, Habitat and Ecology:
New Caledonia, where it is common throughout La Grande-Terre as a semi- emergent to emergent palm in moist forests on a variety of soil types at elevations of 10 to 1200m. The species is most prevalent in the Bourail and Boghen River basin areas. It is a common horticultural subject.

Description: A tall palm with a smooth, evenly ringed, grey trunk to 15m in height and 25cm in diameter. The spreading leaves number about 10. The new leaf is red on exposure. The crownshaft is elongate to 1.2m, blue-green to green with lighter mottling. The petiole, to 45cm long, extends into a rachis to 4m long. The leathery leaflets, deep green above, lighter beneath, number about 38 per side and measure from 1cm to 7.3cm wide and 11cm to 1.5cm long with ramenta present on the mid rib and lateral veins beneath.

The cream/green inflorescence is spreading with rachillae to 50cm long. The peduncular bract is not completely enclosed within the yellowish green prophyll to 85cm long by 16cm wide. Male flowers are pinkish/cream to 2.6cm wide at anthesis, sepals are much shorter than the petals, 40 to 55 stamens. Female flowers are about 1cm high. The red fruit is obovoid to subglobose, to 4.6cm long by 2.6cm in diameter, endocarp lacks an operculum, the seed is sub-globose.

Figure 20: Following page. Recently emerged leaf of *Chambeyronia macrocarpa* (GC).

Figure 21: Opposite following page. Upper left. *Chambeyronia macrocarpa,* New Caledonia (RF). Upper right. *Chambeyronia lepidota* on Mt. Panie' at 600m asl, north-eastern New Caledonia (O). Lower. Infructescence with mature fruit of *Chambeyronia macrocarpa* (RF),

HEDYSCEPE H.A. Wendl. & Drude 1875

A monotypic genus endemic to Lord Howe Island, closely related to *Rhopalostylis* from Norfolk and Raoul Islands and New Zealand. It differs in being protandrous (male flowers shed pollen prior to female receptivity), having triads almost to the end of the rachilla and having more than six stamens.

Hedyscepe canterburyana (C. Moore & F. Muell.) H. Wendl. & Drude 1875
syn. *Kentia canterburyana*

Locale, Habitat and Ecology:

Lord Howe Island on Mts. Lidgbird and Gower, where it occurs as an emergent on steep slopes in moist sub-montane forests on basalt at elevations of 335 to 830m. This is an area where vascular and non-vascular epiphytes are abundant and the soil is constantly moist due to regular rainfall and frequent cloud cover.

Description:

A medium palm with a grey, prominently ringed trunk to 10m in height and 20cm in diameter. The arching, spreading leaves number about 8. A grey-blue crownshaft, to 50cm long, extends into a short petiole and an arching rachis to 3m in length. The erect, linear leaflets number about 30 per side, dark green above, paler beneath and measure to 3cm wide and 30cm long. Scales are present on both the upper and lower ribs.

The spreading inflorescence is infrafoliar, to 40cm long, much branched with fleshy rigid rachillae. Triads are borne throughout, the males being yellow, to 1.4cm wide, with 9 to 12 stamens with a rounded pistillode base. The females are of similar size with overlapping petals becoming valvate at the apices. The fruit on maturity is dull red, obovoid to 5cm long and 4cm in diameter with apical stigmatic remains. The ovoid endocarp is inoperculate, the seed similarly shaped to 4cm long and 2.5cm wide with an elongate hilum.

Notes:

In the restricted areas of the island where it occurs, this palm is one of the dominant plants. The habitat, at 335-830m, is very exposed on steep slopes not less than 15°, and cliffs. The soil is stony, being of basaltic origin and remains constantly moist as a result of the frequent cloud cover, swirling mists and drizzle. Epiphytes, both vascular and non-vascular are abundant and filmy ferns occupy sheltered sites and niches at ground level. Parts of the canopy of the subtropical sub-montane forest is formed by the palm and tree ferns (*Cyathea* spp) which are abundant. The trunks of many of the taller palms are angled away from ground-level due to the subsidence and movement of the soil mass in which they grow.

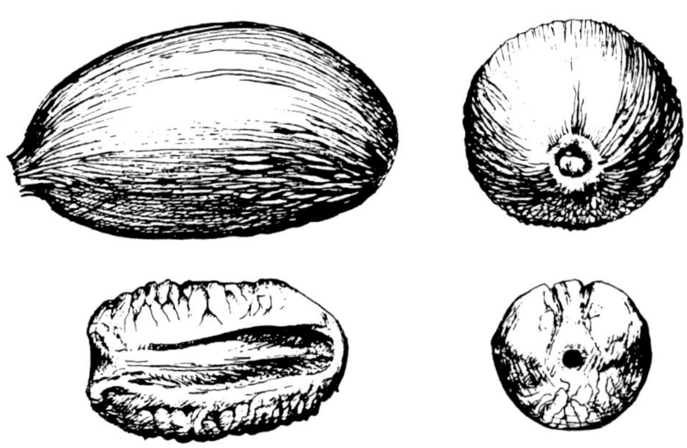

Hedyscepe canterburyana. Endocarp (two views) and seed (two views)(natural size)

Figure 22: Upper. Wind distorted *Rhopalostylis baueri* var. *cheesemanii* above Nash Point on the east coast of Raoul Island (JS).
Lower. Forest canopy formed by *Hedyscepe canterburyana, Bubbia howeana,* and *Dracophyllum fitzgeraldii* on Lord Howe Island (CT).

RHOPALOSTYLIS H.A. Wendl. & Drude 1875

A genus of two species and one sub-species occurring on Norfolk and Raoul Islands, mainland New Zealand and nearby islands and the Chatham Islands, which, at 44° 18'S, is the most southern palm location in the world. The genus is closely related to *Hedyscepe* of Lord Howe Island. *Rhopalostylis baueri* is the type species.

Key to Species

Fruit ovoid to elliptic
  Inflorescence less than 60cm with peduncle less than one-quarter length of the rachis
  Male sepals equal to or longer than the petals
    New Zealand
      R. sapida

  Inflorescence long, exceeding 90cm, with peduncle greater than one-quarter length of the rachis
    Male sepals less than petals in length
      Norfolk Island
        R. baueri var. *baueri*

  Fruit globose
      Raoul Island
        R. baueri var. *cheesemanii*

Rhopalostylis sapida Wendl. & Drude 1878
syn. *Eora sapida*

Locale, Habitat and Ecology:
New Zealand, in coastal areas to as far south as Palm Gully on the Banks Peninsular on the east coast and Greymouth on the west coast of the South Island, the Chatham Islands and most offshore islands off the North Island. This species occurs in high rainfall areas or if in drier areas occupies water courses and moist depressions from sea level to 700m in dense forest as a semi-emergent component. Common in cultivation in subtropical/warm temperate areas.

Description:
A medium palm with a green trunk to 10m in height and 25cm in diameter with closely spaced leaf scars. The bulging crownshaft, occasionally deeply split opposite the petiole, is smooth, green, to 60cm long, extending into a short or absent petiole and erect, peltate scaled, rachis to 5cm long. Leaves number about 14. Leaflets, 80 per side, are closely set and marginally thin, overlapping, acute, dark green above lighter beneath, ascending to 1m long, evenly reducing in length to base and apex. Ramenta is present on the mid-rib beneath and one pair of lateral ribs are prominent above.

The infrafoliar inflorescence, to about 60cm long, is short peduncled, narrowly elliptic in cross section at the base, rachis basally branched and narrowing and becoming terete in cross section toward the apex. Rachillae, to 30cm long and 1.5cm wide, at first cream becoming green, are spreading, flexuous and becoming pendulous in fruit. Prophyll and peduncular bract are of similar size, 30cm long by 15cm wide, smooth, pink to yellowish. Bract fall and male anthesis usually simultaneous. Triads are borne on two-thirds length of the rachilla, closely grouped, paired males distally, buds becoming pink/ lilac. Female receptivity and male anthesis suspected to be infrequently simultaneous but predominantly protandrous. Male flowers are more than twice as long as the female, inserted above the female, stamens 6, sepals equal to or longer than petals. Female flowers with imbricate petals becoming valvate

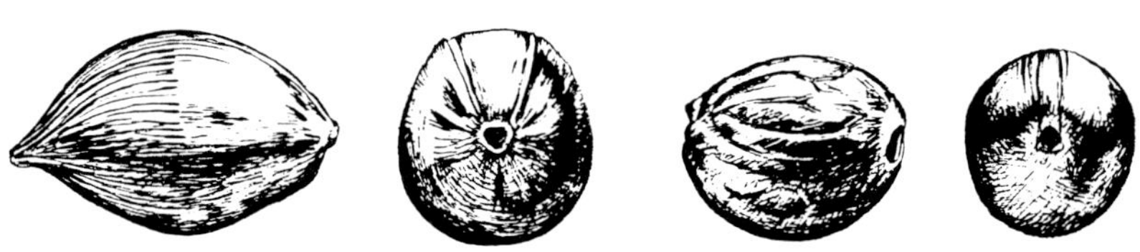

Rhopalostylis sapida Endocarp (two views) and seed (two views)(natural size)

Figure 23: Upper left. *Rhopalostylis sapida,* predominantly in moist habitats, Great Barrier Island, New Zealand (DE).
Upper right. *Rhopalostylis baueri* var. *baueri* in the Norfolk Island National Park (GH).
Lower. Infructescence of *Rhopalostylis baueri* var. *baueri* with maturing and very immature fruit (GH).

at the apices. Fruits, to 1.2cm long by 8mm in diameter, are ovoid to elliptic, brick red at maturity, stigmatic remains apical. Epicarp smooth, mesocarp fibre adherent to the endocarp. Seeds elliptical with an apical hilum and narrow longitudinal raphe.

Rhopalostylis baueri (Hook f.) Wendl. & Drude var. ***baueri*** Sykes
syn. *Rhopalostylis baueri*

Locale, Habitat and Ecology:
Endemic to Norfolk Island in subtropical forest on basalt from sea-level to 319m on Mt. Bates which is the highest point on the island. This species is the visually dominant and major component of 2 recognised forest types. Firstly, Palm/Hardwood Forest, where it occupies the understorey of the continuous canopied forest which is dominated, in size, by hardwood trees such as *Cassine curtipendulum, Nestegis apetala, Rapanea crassifolia* and *Baloghia lucida;* secondly, Palm/Treefern Forest which is confined to narrow strips and gullies. the canopy is markedly broken, the palms along with the treeferns *Cyathea brownii* and *C. australis* var. *norfolkiensis* are semi-emergent. Due to the clearing of the island for settlement, cultivation and pastures, this species is now confined to areas generally above 200m in the Norfolk Island National Park. Only a few palms survive in disturbed forest outside the park area. Green parrots have been observed showing a preference for the fruit of this species. The birds peel the flesh from the fruit, ingest the mesocarp whilst discarding the epicarp and seed. This palm is a successful horticultural subject in mild subtropical/warm temperate areas.

Description:
A tallish palm with a bright green trunk, with raised whitish leafscars, to 15m in height and 20cm in diameter. The occasionally bulging crownshaft is green with a light, brownish tomentum, to 1m long extending into a short petiole and a peltate scaled rachis to 4m long, ascending and erect to horizontal. Leaves number about 14. Leaflets, about 80 per side, darker green above, lighter beneath, are closely set, marginally thin, overlapping, acute, ascending to 1m long at midleaf, evenly reducing in length to the base and apex. Ramenta are present on the mid rib beneath and 1 pair of lateral ribs are prominent above.

The infrafoliar inflorescence, to about 90cm long, has a peduncle about one-third the length of the rachis. Rachillae to 50cm long and 1.4cm wide are at first whitish, becoming green in fruit, flexuous and more or less pendulous. Prophyll and peduncular bract are of similar size, 40cm long by 18cm wide, smooth, pinkish to yellow/green. Triads are borne almost throughout the rachilla, spirally arranged, paired males distally, buds whitish becoming lilac. Male flowers more than twice as long as females, sepals shorter than the petals, stamens 6. Female flowers with basally imbricate petals becoming valvate at the apices. Fruit, to 1.5cm long by 1cm in diameter, are ovoid, elliptic, bright red at maturity, stigmatic remains apical. Seeds elliptical, dull brown with an apical/lateral hilum.

Rhopalostylis baueri (Hook.f.) Wendl. & Drude var. ***cheesemanii*** (Becc. ex Cheesem.) Sykes 1977
syn. *Rhopalostylis cheesemanii*

Locale, Habitat and Ecology:
Endemic to Raoul Island of the Kermadec Islands, abundant in the eastern wet forests particularly on the higher ridges as an understorey component in forest dominated by *Metrosideros kermadecensis* or other large trees, or as scattered individuals on the drier ridges towards the headlands. Rare or absent elsewhere on the island:

Description: A tallish palm to 15m in height and 20cm in diameter with closely spaced leaf scars on a grey/green trunk, The bulging crownshaft, occasionally split opposite the petiole, is green, smooth,to 70cm long extending into a very short petiole and erect peltate scaled rachis to 4m long. Leaves number about 14. Leaflets, about 80 per side are closely set and marginally thin, overlapping, acute, dark green above, lighter beneath to 1.4m long at mid leaf, evenly reducing in length to the base and apex. Ramenta is present on the mid rib beneath, one pair of lateral ribs prominent above.

The infrafoliar inflorescence, to about 90cm long, is short pedunculate. Rachis is basally branched. Rachillae, to 40cm long and 1.5cm in diameter at first whitish, then becoming light green in fruit, flexuous and pendulous, remaining white at male anthesis. Prophyll and peduncular bract are of equal size,

Figure 24:*Rhopalostylis baueri* var. *cheesemanii,* flowering and with young fruit, a common sight on Raoul Island in December to February period (BDS, NZGS).

40cm long by 18cm wide, smooth pink to yellowish/green. Triads are borne on two-thirds the length of the rachilla, paired males distally, whitish becoming pink. Male flowers more than twice as long as the females, sepals less than petals in length, stamens 6. Female flowers with basally imbricate petals becoming valvate at the apices. Fruit, globose to sub-globose, to 1.6cm long by 1.35cm in diameter, brick red at maturity, stigmatic remains apical. Epicarp smooth, endocarp with tightly adherent fibres with vascular strands curving obliquely from the hilum. Seed is sub-globose, with a broad longitudinal hilum.

Notes:

Although the genus is reported to be protogynous (i.e. female flowers become receptive prior to male anthesis) (Uhl & Dransfield 1987), there is only scant evidence to support this as the norm. Rather, the flowering mode is suggested to be protandrous, with only apparent protogynity through the simultaneous male anthesis/female receptivity at the time of bract release. It is reported (Elser 1969) that mature inflorescences are sometimes 'captured' by tenacious leaves which have resisted falling and thus release the mature inflorescence. It is possible that such inflorescences may retain the shed pollen within the bracts until female receptivity which occurs about 14 days after anthesis. More study needs to be undertaken to clarify the flowering sequence of this genus, and, if any, the differences between the species and related taxa.

The length of the male sepals in relationship to the petals is also more variable between species than suggested in the descriptive literature.

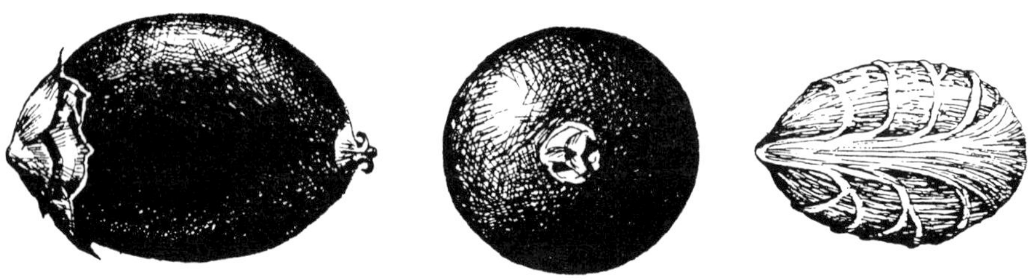

Rhopalostylis baueri var. *cheesemanii.* Fruit (two views) and endocarp (three times natural size)

ACTINOKENTIA Dammer 1906

A genus of two species endemic to New Caledonia. Relationships to other genera are obscure but *Kentiopsis* appears close. *Actinokentia divaricata* is the type species.

Key to Species

Petiole long, terete in cross section
    Floral bracts not as high as flowers
        Male flowers sepals only ½ as high as petals
        Fruit 2.4cm by 9mm — *A. divaricata*

Petiole short, channelled above
    Floral bracts exceeding flowers in height
        Male flowers sepals and petals of equal length
        Fruit 3cm by 1.3cm — *A. huerlimannii*

Actinokentia divaricata (Brongn.) Dammer 1906

syn. *Kentiopsis divaricata*
Kentia polystemon
Actinokentia schlechteri

Locale, Habitat and Ecology:

New Caledonia, where it is commonly found in the southern half of La Grande- Terre on serpentine soils in dense forests at 200-700m as an understorey component. The habitat receives a high annual rainfall. The associated forests are very dense with under shrubs and small trees. Other palms which are found nearby include *Cyphokentia macrostachya, Basselinia gracilis, B. pancheri* and *Burretiokentia vieillardii.*

Description:
A small to medium palm to 7m tall with a slender trunk to 8cm in diameter. Nodes are not prominent. Roots are adventitious and prickly. Leaves number from 4 to 6, spreading, ascending to horizontally held with a long reddish/ pink petiole to 85cm extending into an arching rachis to 1.45m long. The crownshaft is glossy brown/green, elongate, convex in profile, to 65cm long. Leaflets number about 25 per side, dark green above, paler beneath with ramenta on the mid rib beneath. They are widely spaced, linear-acute and twisted laterally, from 47cm to 96cm long and 1.4cm to 6.2cm wide. The emerging leaf is red/crimson.

The glabrous infrafoliar inflorescence spreads divaricately to 40cm long and wide. Ultimate rachillae are to 18cm long. The lepidote prophyll, to about 23cm long, encloses the peduncular bract to about 22cm long. Triads are borne on two-thirds the length of the rachilla, subtended by prominent bracts. Male flowers are dull red in bud, about 5.5mm long, with the pistillode equal the length of the stamens which number about 20. Female flowers to 6mm are pinkish/brown. The ellipsoidal fruit, to 2.5cm long by 1cm in diameter, have apical stigmatic remains and large persistent petals and are dull purple on maturity. The endocarp is striately patterned, the seed has oblique grooves radiating from the elongate hilum.

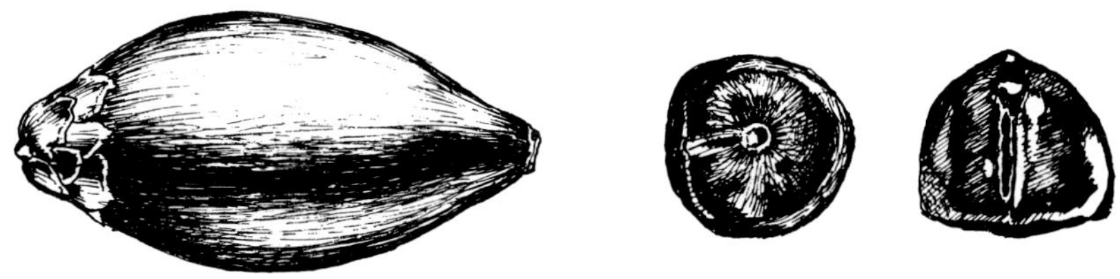

Actinokentia divaricata. Fruit and encocarp (two views)(natural size)

Actinokentia huerlimannii H.E. Moore 1980

Locale, Habitat and Ecology:
Endemic to New Caledonia where it is of limited occurrence on Mt. Nekando and Haute Ni in the southern and central sectors of La Grande-Terre respectively, at an altitude of 850-1000m. It is an understorey palm in moist forest on serpentine soils in association with ultrabasic rocks which are peridotitic in nature. Other palms which share the moist, cool habitat and grow within the vicinity include *B. humboldtiana* and *B. porphyrea,* at least on Mt. Nekando.

Description:
A small to medium palm with a lightly ringed trunk to 9cm in diameter, without exposed adventitious roots. Leaves, to about 1.1m long, number 5, spreading and ascending with a short, reddish petiole. The crownshaft is olive green. Leaflets number about 20 per side, glossy green above, paler beneath, with prominent ramenta on the mid rib beneath. Leaflets measure from 15cm to 56cm in length and from 4mm to 2.5cm in width.

The lepidote infrafoliar inflorescence is to 40cm long, short pedunculate with rachillae to 22cm long. Branch and rachilla bracts are prominent. Acute bracts to 6mm long subtend triads which are spirally arranged. Male flowers, to 4mm high have about 20 stamens. Females are about 7mm long with brown scaled sepals and petals. The fruit is ellipsoidal, 3cm long and 1.3cm in diameter. Retained petals at the base are 7mm long.

KENTIOPSIS Brongn. 1873

A monotypic genus endemic to New Caledonia. It is closely related to *Mackeea* also of New Caledonia.

Kentiopsis oliviformis (Brongn. & Gris) Brongn. 1873
syn. *Kentia oliviformis*

Locale, Habitat and Ecology:
New Caledonia, where it occurs rarely in the central sector of La Grande-Terre in the vicinity of Bourail, Kouaoua and the Boghen Valley in the west and Canala in the east, as an emergent species in

moist forest on acid rocks and basalt at an elevation of 30 to 300m. Uncommon in cultivation.

Description:
A tall palm to a height of 30m with a smooth but ringed grey trunk to a diameter of 30cm with a thickened base. The leaves, numbering about 9, have irregular phyllotaxy. The crownshaft, to 1.1m long, swollen with a symmetrical profile, is green with pale scales becoming glabrous. The petiole is very short to 10cm in length, extending into a horizontally held rachis to 3m. The leaflets, about 55 per side, are glossy dark green above, paler beneath with membranous ramenta on the mid rib beneath. They measure from 2.5cm to 5.5cm wide and from 19cm to 94cm long, the basal leaflets forming lorae.

The short pedunculate inflorescence is covered in scales. The yellowish prophyll, to 68cm long and 21cm wide, encloses the peduncular bract, both of which are covered in pale scales. The rachis is branched at the base, with glabrous, thin, rigid rachillae to 30cm long. Triads are borne throughout the rachilla, the males to 5.5mm high with the pistillode equal to the stamens in length and about 35 stamens, the females to 7mm high. The fruit is ellipsoidal, about 2cm long and 1cm in diameter, with apical stigmatic remains. The inoperculate endocarp encloses a curved ovoid seed to 1cm long and 4mm wide with an elongate hilum.

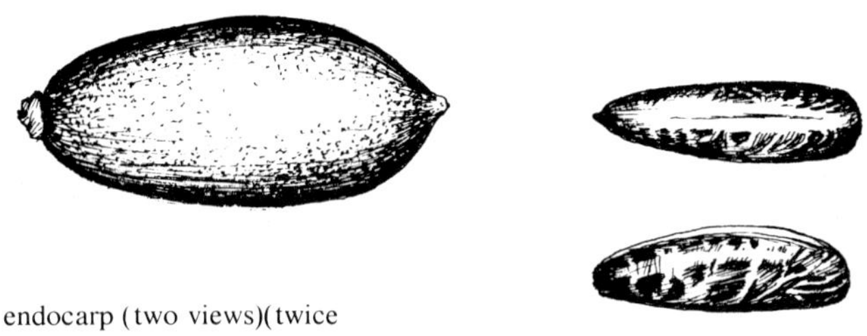

Kentiopsis oliviformis. Fruit and endocarp (two views)(twice natural size)

MACKEEA H.E. Moore 1978

A monotypic genus endemic to New Caledonia. It is closely related to *Kentiopsis,* also of New Caledonia.

Mackeea magnifica H.E. Moore 1978

Locale, Habitat and Ecology:
New Caledonia, in the north-west sector of La Grande-Terre, in moist montane forests and residual forest on stream edges at 500 to 700m, particularly common in the Haute Mayavetch where it grows with *Cyphophoenix elegans.* Other populations are found on the ridges above Diahot Estuary, in the Thiahate Valley and the eastern slopes of Mt. Panie. It is a tall emergent palm readily distinguishable from *C. elegans* by its large size and leaves which are spreading and horizontal, not arching.

Description:
A tall palm to 25m with a grey/brown trunk to 25cm in diameter. Leaves, numbering about 9, are spreading and held horizontally. The crownshaft is prominent, to 1.5m long, green with grey scales extending into a petiole to 20cm long with white scales and a rachis to 2.25m long similarly scaled. Leaflets, about 55 per side, are glossy dark green on both sides, spreading, with brown ramenta on the mid rib beneath, measuring from 1cm to 15cm wide and 27cm to 1m long. The basal leaflets extend into lorae.

The infrafoliar inflorescence, to 55cm long, is grey/green, purplish/brown to pale green, short pedunculate with lepidote tomentum, thickest between the bracts. The rachis, with peduncular base, to 18cm long, becoming glabrous, has ultimate rachillae to 44cm long. Brownish bracts subtend the branches. Prophyll and peduncular bract, to 55cm long, are covered in tomentum. Triads are spirally arranged. Male flowers are brown in bud, white and widely spreading at anthesis with a very short pistillode and about 35 stamens. Female flowers are much smaller than the males with sepals 5mm high and 8mm wide. Fruit, to about 2.2cm long and 1.2cm wide, is ellipsoidal, with apical stigmatic remains, dark green at maturity and a persistent brown perianth. Fibres within the mesocarp are basally adherent to the endocarp, which is vitreous and inoperculate, with a longitudinal groove and slightly depressed base. The seed is ellipsoidal, 1.6cm long by 9mm in diameter, with a linear hilum.

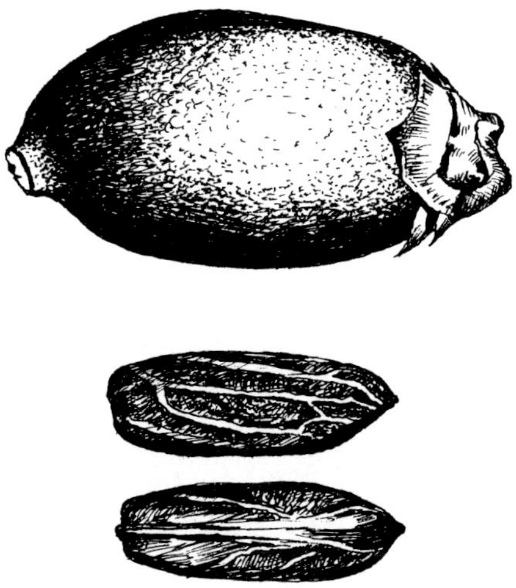

Mackeea magnifica. Fruit and endocarp (two views)(twice natural size)

Figure 25: Following page. *Actinokentia divaricata* in low altitude forest in the middle reaches of Riviere Bleue between the refuge hut and Port Germain, south New Caledonia (AD).

Figure 26: Opposite following page. Vividly coloured leaf of *Actinokentia divaricata* Mount Koghi area, south New Caledonia (AD)

LINOSPADICINAE
J.D. Hooker
Subfamily: Arecoideae
Tribe: Areceae

Calyptrocalyx
Laccospadix
Linospadix
*Howea**
* Indigenous genus

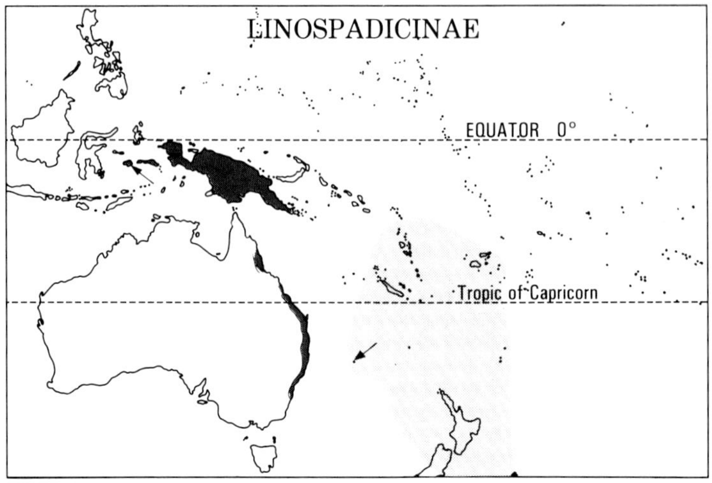

Description:
Small to tall, understorey or emergent; solitary or clump-forming palms; leaf non-sheathing, simple pinnately veined or pinnately leaved; unarmed; monoecious, pleonanthic. Inflorescence a single spike or multiple spikes on united peduncles; unisexual flowers positioned in sunken pits, fibres not attached to the endocarp, lacking an operculum.

Distribution:
From equatorial to warm temperate; sea-level to moderate altitudes; rainforest, moist forests, swamp forests, coastal forests. East Malesia, New Guinea, eastern Australia and Lord Howe Island.

Notes:
By criteria established by recent studies (Moore & Uhl 1982), this is an advanced SUBTRIBE. The inflorescence, which is either a single spike or two or more spikes emanating from united peduncles and sharing the one prophyll, is considered an advancement and a major trend of inflorescence development within ARECACEAE. *Howea* is similar to *Calyptrocalyx* which is confined to Malesia/New Guinea. The genera which are geographical intervenors are *Laccospadix*, which is monotypic and restricted to north-eastern Australia and *Linospadix*, which is well developed and found in both New Guinea and Australia.

HOWEA Becc. 1877

A genus of two species endemic to Lord Howe Island, most closely related to *Laccospadix* and *Linospadix* of eastern Australia. *Howea belmoreana* is the type species.

Key to Species

Leaves arching with erect leaflets
Inflorescence a single spike — *H. belmoreana*

Leaves horizontal with drooping leaflets
Inflorescences multiple — *H. forsteriana*

Howea belmoreana (C. Moore & F.Muell) Becc. 1877
syn. *Kentia belmoreana*
Grisebachia belmoreana

Locale, Habitat and Ecology:
Endemic to Lord Howe Island, where it is abundant on slopes in forests on basaltic soils at 5-250m. It is an emergent or semi-emergent species, depending upon the stature of the trees with which it grows. Its range occasionally overlaps with that of *H. forsteriana.*

Description:
A tall palm to 12m with a green trunk becoming grey to 15cm in diameter, with prominent white rings, expanded at the base. The erect, arching leaves number about 18. Yellowish/green clasping leafbases are covered in dried meshed fibre, the petiole is arching to 50cm and extending into a rachis to 4m long. The linear acute leaflets are held erect, becoming horizontal only toward the apex, and measure to 3cm wide and 60cm long.

The inflorescence is a simple, curved spike becoming pendulous, to 1.5m long and to 2.5cm in diameter, initiated within the axils of the leaves, but later being displayed below the leaves once they have fallen. The woody prophyll, to 18cm long, and the papery peduncular bract, attached toward the apex of the prophyll, are persistent and slow to decay. Triads are borne in recesses, spirally arranged throughout the entire length, subtended by broad bracts. This is a protandrous species, 12 months passing between male anthesis and female receptivity. Male flowers have woody petals, valvate, with 30-70 stamens and lacking a pistillode, persisting in the dried state. Female flowers are smaller than the males, reddish in bud with ciliate sepals. The fruits are ellipsoidal, attenuated toward the apex and the base, to 4cm long, with apical stigmatic remains and becoming brownish/red at maturity. The endocarp is brittle, not adherent to the seed, ellipsoidal with a lateral ridge. The seed has a lateral hilum and basal embryo.

Notes:
This species is generally confined to slopes situated between 5-250m on stony basaltic soils in subtropical lowland forest which has a canopy height not exceeding 12m. In most of its habitat, the palms form the canopy, which is sparser than that formed by *H. forsteriana,* but still dense enough not to allow strong light to penetrate to the soil level. Small trees, such as *Bubbia howeana* and *Randia stipulosa,* herbs and palm seedlings of various heights form most of the vegetation at ground level. In the more heavily forested areas which contain *Ficus columnaris* and other large trees, this species forms a conspicuous understorey component.

Of the total area on the island occupied by this species and *H. forsteriana,* about 40% could be considered to be dominated by this species. The two species merge in the areas around the north-west and north-east lower slopes of Mt. Lidgbird and the north-western lower slopes of Mt. Gower and to a lesser extent on the southern slopes of Settlement Hill on the north of the island.

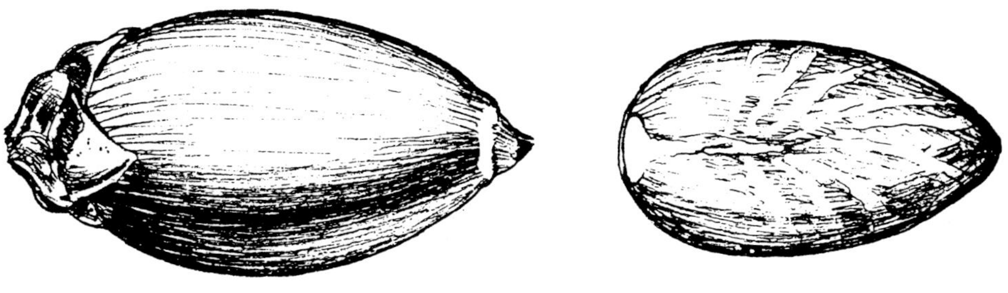

Howea belmoreana. Fruit and endocarp (twice natural size)

Howea forsteriana (C. Moore & F. Muell) Becc. 1877
syn. *Kentia forsteriana*
Grisebachia forsteriana
Denea forsteriana

Locale, Habitat and Ecology:
Endemic to Lord Howe Island, where it is a dominant lowland species occupying flat or gently sloping areas on soils derived from calcarenite at 0-360m.. It is an emergent species often found in exposed sites within metres of the sea-shore and in dense stands.

Description:
A tall palm to 25m with a prominently ringed green trunk to 15cm in diameter, expanded at the base. The horizontal to pendulous leaves number about 15. The clasping leafbases are covered in dried

meshed fibre, the petiole to 30cm extends into a horizontally held to pendulous rachis to 4m long. The linear acute dark green leaflets are held horizontal or pendulous, number to about 70 per side, with a prominent whitish/yellow mid rib and measuring to 3cm wide and 60cm long.

The inflorescence consists of multiple, pendulous spikes, 3 to 5, fused at the base into a broadish peduncle, initiated and arising from within the leafbases but later displayed below as the leaves fall. The woody prophyll, to 18cm long, encloses the inflorescence bases but each individual spike has its own papery peduncular bract, attached at about the apex of the propyhll. All bracts are persistent and slow to decay. Triads are borne in recesses spirally arranged throughout the entire inflorescence, subtended by prominent bracts. Male flowers have woody petals, are valvate, with 30-70 stamens and lacking a pistillode, persisting in the dried state. Female flowers are smaller than the males, reddish in bud with ciliate sepals. The fruits are ellipsoidal, attenuated at the apex, 5cm long and 2cm in diameter, with apical stigmatic remains, becoming red at maturity. The endocarp is brittle, not adherent to the seed, ellipsoidal with a lateral ridge. The seed has a lateral hilum and basal embryo.

Notes:

This palm is the most dominant plant species on Lord Howe Island, being found at altitude from 0-360m. It forms almost pure stands in subtropical lowland forest which has a canopy height of 10-20m out of which the palms may emerge to 15- 25m. The habitat is flat or gently sloping and the density of the palms is estimated to be 1770 per hectare (Pickard, 1983). Other plants which grow within this palm forest include the large *Ficus columnaris, Pandanus forsteri, Randia stipulosa, Flagellaria indica* and *Smilax australis.* Vascular epiphytes are uncommon. Only in a limited number of locations does this species and *H. belmoreana* co-exist, these being listed in notes under that species. These sites are where *H. forsteriana* strays from its preferred calcarenite sandy subsoils and intrudes onto the slopes of predominantly basaltic origins. This genus presently dominates about 16% of the island's 1520 hectare land surface.

Figure 27: Upper left. *Howea belmoreana* forest on Lord Howe Island, with Helen Tiffin (CT).
Upper right. *Howea forsteriana* on Lord Howe Island (CT).
Lower. Infructescence with maturing fruit of *Howea belmoreana.*

PTYCHOSPERMATINAE

J.D. Hooker
Subfamily: Arecoideae
Tribe: Areceae

Drymophloeus
*Balaka**
Ptychosperma
Carpentaria
Normanbya
Ptychococcus
*Veitchia**
Wodyetia
Brassiophoenix
* Indigenous genera

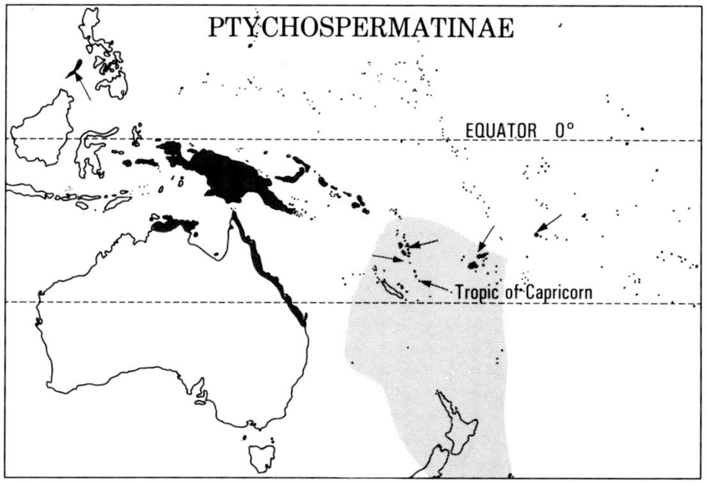

Description:
Small to tall, understorey to emergent; solitary or clump-forming palms; leaf sheathing, pinnate, single ribbed leaflets; unarmed; monoecious, pleonanthic; stamens numerous; endocarp lacking an operculum.

Distribution:
From equatorial to sub-tropical; sea-level to moderate altitudes; rainforest, moist forests, swamp forests, dry vine forests, monsoon forests, coastal forests; East Malesia, New Guinea, Philippines, Solomons, northern Australia, Vanuatu, Fiji and Samoa.

Notes:
Veitchia and *Balaka* are well developed totalling at least 20 species between them. *Veitchia* may be considered the most successful genus in the region being found on all the large islands and many small islands of Vanuatu and Fiji. The SUBTRIBE is best developed in New Guinea and the Solomons whilst three monotypic genera, *Carpentaria, Normanbya* and *Wodyetia,* are found in tropical Australia. It is also suggested by fruit and seed characters, that migration and development of the SUBTRIBE through the South-West Pacific proceeded through the Solomons to Fiji and thence to Vanuatu. Fruits and seeds of *Balaka* and some Fijian *Veitchia* are similar to *Ptychosperma* whilst those taxa from the SUBTRIBE in Vanuatu are dissimilar, displaying seeds which are only terete in cross-section.

VEITCHIA H.A. Wendl. 1868

A genus of about 16 species occurring in Vanuatu, Fiji and Philippine Islands. Twelve species are annotated in this work, six in Vanuatu and six in Fiji. A number of taxa are known from fragmentary collections and are most likely synonymous with the known species.Comprehensive collections need to be made from the more remote areas to clarify the status of some species. Closely related genera include *Ptychosperma* and *Carpentaria.*
Veitchia joannis is the type species.

Figure 28: Leaf of *Veitchia simulans* from Taveuni, Fiji.

Key to the Species

Leaflets densely lepidote beneath
Inflorescence glabrous
Rachis not recurved
Rachillae rigid
Fruit oblong to ellipsoidal
Stamens around 100
Fruit to 4.5cm long by 2.4cm in diameter
Efate, Vanuatu
V. montgomeryana

Fruit to 3cm long and 1.7cm in diameter
Espiritu Santo, Vanuatu
V. macdanielsii

Stamens around 50
Fruit less than 2cm long
Malekula, Vanuatu
V. winin

Fruit subglobose to broadly ovoid
Taveuni, Fiji
V. simulans

Rachillae flexuous
Fruit to 4cm long and 3cm in diameter
Aneityum, Vanuatu
V. spiralis

Leaflets not or only sparsely lepidote beneath
Rachillae rigid
Fruit oblong to ellipsoidal
Fruit less than 2cm long
Leaflets bluntly truncated
Viti Levu, Kandavu, Fiji
V. vitiensis

Rachillae flexuous
Leaflets not bluntly truncated
Fruit 4cm long by 2cm in diameter
Vanuatu,
V. arecina

Fruit less than 2cm long
Vanua Levu, Fiji,
V. pedionoma

Inflorescence lepidote-floccose
Rachis recurved
Petiole long
Leaflets closely spaced
Vanua Levu, Taveuni, Fiji
V. petiolata

Leaflets not closely spaced
Leaflets apically broad
Vanua Levu, Fiji
V. sessilifolia

Petiole not long
Stamens around 100
Fruit greater than 5cm long
Leaflets narrowing toward the apex
Fiji
V. joannis

N.B. This key excludes *V. metiti,* an incompletely known species from Vanua Lava in Vanuatu.

Figure 29: Towering group of *Veitchia montgomeryana,* near Port Vila Airport, Vanuatu.

Veitchia montgomeryana H.E. Moore 1957

Locale, Habitat and Ecology:

Vanuatu, endemic to Efate, as an emergent at low altitudes in moist forests, usually in shallow soils overlying well drained coral substratum, but with some populations on deep alluvial soils close to rivers and streams especially in their lower reaches. This species is now becoming rare and endangered due to the collection of the palm heart which is used as a salad vegetable by European restauranteurs in resorts in Port Vila and Manuro.

Description:

A tall palm to 30m with a grey trunk to 20cm in diameter, expanded at the base with a mound of adventitious roots. The internodes are elongate, grey/green on the newly exposed upper trunk, the nodes narrow and lighter grey. Leaves number about 10, ascending to horizontal with slightly arching apex. The crownshaft is elongate, to 1.25m long, narrowing toward the petioles, green with grey floccose scales, dark brown scales becoming very dense below and into the petiole which is about 25cm long, orangish above, scaled below, the rachis to 2.6m long is similarly scaled. Leaflets, about 60 per side, are held horizontally to slightly pendulous, the apex obliquely truncate praemorse, the upper margin longer, dark green and glabrous above, densely lepidote beneath, particularly at the base. The mid rib is prominent, raised above, leaf margins thickened, more so basally, all with brown scales below. Leaflets from 32cm to 80cm long and 1.5cm by 7cm wide, the lower-most reduced to lorae about 1m long.

The infrafoliar inflorescence is broader than long, 60cm long by 80cm wide, stout, yellowish in fruit, glabrous, twice branched, rachillae rigid, ultimately to 25cm long, thin, narrowing at apex. The peduncle is stout, broadly elliptical in cross- section, the rachis to 50cm long and the rachillae are angled. The prophyll is densely scaled, light yellowish green/grey, the peduncular bract similarly shaped and coloured but less densely scaled, both early deciduous. Triads are borne almost throughout, paired males distally, bracts inconspicuous. Male flowers are bullet shaped in bud, green, glabrous, about 1cm high, 7mm wide, sepals about one-quarter the length of the petals which are valvate, thickish, widely spreading at anthesis, stamens about 100-130, anthers white, pistillode longer than the stamens, bottle shaped, bifid at the apex. Female flowers are rounded, to 1cm in diameter. Fruit, bright red at maturity, is oblong, about 4.5cm long by 2.4cm in diameter, tapered toward the base, with a cupular persistent perianth. The epicarp smooth, the mesocarp to 4mm thick, thinly fibrous, the endocarp is thin, adherent to the oblong/ellipsoidal seed which is slightly depressed on the side of the elongate hilum, the raphe has loose descending fibres, the embryo is basal.

Veitchia macdanielsii H.E. Moore 1957

Locale, Habitat and Ecology:

Vanuatu, endemic to Espiritu Santo, particularly in the south and east at low to moderate altitudes, as an emergent in coastal forests in soils overlying coral substratum and on well drained slopes of volcanic origin. It is incorrectly reported as being indigenous to New Caledonia where it is known only in cultivation.

Description:

A tall palm to 25m with a grey trunk to 20cm in diameter, expanded at the base. Internodes are elongate, grey/green on the newly exposed upper trunk. Leaves number about 10, ascending, horizontal to arching. The crownshaft is elongate to 1m, lepidote, with dark scales becoming denser toward the petiole. The petiole is short, lepidote above and below, the rachis at first similarly scaled but becoming glabrous toward the apex. Leaflets, about 55 per side, are slightly pendulous to pendulous, the apex obliquely truncate, with the upper margin longer, praemorse, dark green and glabrous above, densely lepidote beneath, more so toward the base. The mid rib is prominent above, with scales below, the marginal veins thickish. Leaflets are from 1.6cm to 7cm wide by 15cm to 1m long, the lower leaflets reduced to lorae which are inserted at right angles to the rachis, directed forward and becoming pendulous toward mid-leaflet.

Figure 30: Upper left. Infructescence with mature fruit of *Veitchia montgomeryana* held by Chanel Sam, ORSTOM, Port Vila, Vanuatu.
Upper right. Peduncular bract of *Veitchia montgomeryana.*
Lower. Mature fruit of *Veitchia montgomeryana.*

The infrafoliar inflorescence is whitish in bud and fruit, about 1.2m long and 90cm broad, thrice branched with rigid rachillae to 5mm thick. The prophyll and peduncular bract are pale lepidote-tomentose, of similar shape, about 60cm long and 14cm wide. The peduncle is stout, broadly elliptical in cross-section, the rachis and rachillae angled, at first with pale scales, these deciduous, becoming glabrous. Triads are borne nearly throughout, with paired males distally, bracts inconspicuous. Male flowers are bullet shaped in bud, glabrous, about 1.5cm high, 7mm in diameter, with sepals less than the length of the petals, widely spreading at anthesis, with about 100 stamens with white anthers, and an elongate pistillode. Female flowers are rounded to 8mm in diameter. Fruit is ellipsoidal, bright red at maturity, about 3cm long and 1.7cm in diameter, with stigmatic remains subapical, attenuated to a point 3mm long and wide. The epicarp is smooth, mesocarp to 2mm thick, fibrous, the endocarp fragile. The seed is oblong-ellipsoidal to 2cm long by 1.4cm in diameter, impressed at the base, with an apical raphe with descending branches. The embryo is basal.

Veitchia macdanielsii. Fruit and endocarp (twice natural size)

Veitchia winin H.E. Moore 1957

Locale, Habitat and Ecology:
Vanuatu, most commonly on Malekula, where it grows in moist forest from sea- level to about 550m. On Espiritu Santo it has been observed growing at low altitude in volcanic alluvium and on Erromango in the vicinity of the Ipota River and Pentecote from 80-550m. This species dominates much of the primary and secondary forests on Malekula, an island little disturbed by Melanesian man, but now under increasing pressure from expanding coconut plantations and logging operations.

It is a successful horticultural introduction.

Description:
A tall palm to 20m with a grey trunk to 17cm in diameter, enlarged at the base. The internodes are elongate, grey/green lepidote on the newly exposed upper trunk. Leaves, about 10, are horizontal to arching. The crownshaft is elongate to 1m, green with whitish lepidote-tomentum, red-brown scales becoming dense toward and into the short reddish/brown petiole, the rachis to about 2.8m similarly scaled below, glabrous above. Leaflets, about 50 per side, are pendulous, acuminate to obliquely truncate, praemorse, glabrous, dark green above, lepidote beneath, more so basally, the mid rib prominent, with brown scales on the mid rib and thickened marginal veins below. Leaflets 20cm to 75cm long and 1.3cm to 11cm wide, the lower-most reduced to narrow lorae.

The infrafoliar inflorescence, to 60cm long and 80cm wide, is glabrous, yellowish in fruit, 3 to 4 times branched, ultimate rachillae rigid, thin, to 20cm long. The prophyll is densely white tomentose, to about 45cm long and 14cm wide, the enclosed peduncular bract similar but slightly smaller. The peduncle, to 15cm long, is densely white lepidote-tomentose, broadly elliptical in cross-section, the rachis short. Triads are borne almost throughout, sub-distichously arranged paired males distally, bracts inconspicuous. Male flowers are whitish, about 1cm high, with valvate petals, sepals one-third their length, stamens around 50, pistillode attenuated with a trifid apex, exceeding the stamens in length. Female flowers are whitish, rounded to 5.5mm high and 4.5mm in diameter. The fruit is ellipsoidal, bright red at maturity, about 1.8cm long by 9mm in diameter, perianth persistent, stigmatic remains

Figure 31: *Veitchia macdanielsii* in lowland forest, south Espiritu Santo, Vanuatu.

sub-apical, to 2mm long. The epicarp is smooth, the mesocarp fibrous, the endocarp thin and fragile, not adherent to the seed. The seed is ovoid with a depressed base, 1cm long by 7mm in diameter, an elongate hilum and raphe branches descending from the apex. The embryo is basal.

Veitchia winin. Fruit and endocarp (two views)(twice natural size)

Veitchia simulans H.E. Moore 1957

syn. *Veitchia subglobosa*

Locale, Habitat and Ecology:

Fiji, endemic to Taveuni on the north-western slopes, in very moist, dense forests on volcanic soil at 300-600m altitude. This area has a rainfall exceeding 6000mm per annum, falling almost daily for a greater part of the year and frequently during the drier winter months. The soils in the area are very shallow, overlying basaltic rocks. The altitudinal stratification of the individual palm species in this area is precise. This species occupies a 300m range along the slopes which are both steep and gentle, whilst *Clinostigma exorrhizum* begins at about 300m extending to the summits of the peaks in the centre of island. In the lower zones is found *Calamus vitiensis* from about 350m to almost sea level. This species appears more closely related to the *Veitchia* in Vanuatu. and even to *Carpentaria* of northern Australia than to the taxa on Taveuni and elsewhere in Fiji.

Description:

A medium palm to about 7m with a closely scarred, grey/green trunk to 15cm in diameter and an expanded base. Leaves, numbering about 8, are horizontally held to about 3m long. The crownshaft is dark green, to 40cm long with grey scales, denser toward the petiole which is to 25cm long, 4cm wide at the base, extending into a horizontal, slightly arching rachis, the surfaces of which are densely lepidote above and below. The leaflets are semi-pendulous, about 25 per side, with a prominent mid rib, glossy green above, densely red/brown lepidote below, more so toward the base and on the margins. The leaflet apex is obliquely truncate, barely praemorse, tapering from the middle to the apex and the base, the margins thickened and slightly curved. They measure from 36cm to 61cm long and from 2cm to 5cm wide, the lower leaflets reduced to lorae.

The infrafoliar inflorescence, to 55cm long, is held erect to horizontal, rigid, glabrous, greenish/yellow in fruit, three times branched. Branches and rachillae reduce in length toward the apex, the bracts subtending the lower and mid branches are acuminate, absent distally. The ultimate rachillae are thickish to 12cm long. Triads are spirally arranged almost throughout, paired males distally, bracts inconspicuous. male flowers are whitish, about 6mm high, petals widely spreading at anthesis, stamens around 50, pistillode exceeding the stamens in length. Female flowers about 1.3cm wide. The fruit is broadly ovoid, slightly attenuated toward the apical stigmatic remains, orange/red at maturity, about 2cm long by 1.7cm in diameter. The epicarp is smooth, the mesocarp thickly fibrous, the endocarp thin and fragile. The seed is ovoid with an elongate raised hilum, 1.3cm long by 1cm in diameter, raphe branches descending from the apex. The embryo is basal.

Veitchia spiralis H. Wendl 1868

syn. *Kajewskia aneityensis*

Locale, Habitat and Ecology:

Vanuatu, endemic to Aneityum, in lowland areas in light forest as an emergent species. Aneityum is a volcanic island with a high rainfall and prone to almost annual incursions of cyclones due to its geographical position. The type locality of *Carpoxylon,* an unrelated rare genus, is also Aneityum.

Figure 32: *Veitchia winin,* an unusual occurrence in central Espiritu Santo, Vanuatu.

Description:

A tall palm to 20m high with a trunk diameter of 20cm, expanded at the base. Leaves number about 11, ascending to horizontal and arching, the apex not pendulous. The crownshaft is elongate, narrowing toward the petioles, densely white lepidote scaled, dark scales denser toward and into the petiole, the rachis similar below but less so above except at the base of the leaflets. Leaflets, about 50 per side, are held horizontal to semi-pendulous, glossy/glabrous above, tapering from the middle to the base and the apex, obliquely truncate and praemorse, minutely lepidote below, the mid rib and marginal nerves with brown scales. Leaflets are about 66cm long by 2.5cm to 8cm wide.

The infrafoliar inflorescence is densely lepidote, branched to 4 orders with angled rachis and rachillae which are ultimately to 14cm long. Triads are borne on the basal one-third, paired or solitary males distal, subdistichously arranged. Male flowers, about 1.5cm high, have sepals about the length of the petals, stamens around 118. The female flowers are to 9mm high and 1cm in diameter. The fruit is broadly ovoid, 4cm long by 3cm in diameter. The mesocarp has dense longitudinal fibres. The seed is subglobose, 2.6cm long by 2.4cm wide, attenuated at the apex, the raphe apical with parallel descending fibres.

Veitchia vitiensis (H. Wendl) H.E. Moore 1957

syn. *Ptychosperma vitiense*
Vitiphoenix vitiensis
Veitchia vitiensis var. *vitiensis*
V. vitiensis var. *parhamiorum*
V. vitiensis var. *microcarpa*
Vitiphoenix smithii
Veitchia smithii

Locale, Habitat and Ecology:

Fiji, widespread but uncommon on Viti Levu, Kandavu and Ovalau, as an understorey to semi-emergent species in sheltered sites in moist forests on volcanic soils from sea-level to 900m. This is a variable species previously considered to be represented by a number of varieties distinguished by leaflet width and length, obliqueness and width of the truncate apex, branching of the inflorescence, fruit size and shape and their positioning on the rachilla. The fruits are consumed by fruit bats (R.H. Phillips, pers.comm.).

Description:

A medium palm, to 16m high with a grey/brown trunk to 20cm in diameter, usually much less and not expanded at the base. Internodes are elongate, the nodes raised. Leaves number about 8, are ascending to arching with a pendulous apex. The crownshaft is moderate, to 45cm long, green with horizontal to oblique patches of dark green lepides becoming dense toward the petiole which is thin, to 50cm long, densely scaled above and below, the rachis arching and similarly scaled below, glabrous above. Leaflets, about 40 per side, are held erect and stiff, or pendulous, tapering from the middle to the base and the apex, obliquely truncate, usually praemorse, glossy above, glabrous below, mid rib prominent with red/brown ramenta below and minute scales on the mid rib and marginal veins near the base. Leaflets measure from 12cm to 45cm long and from 3cm to 6.5cm wide.

The infrafoliar inflorescence, to 1m long, is glabrous, branched to 4 orders, rigid, ultimate rachillae thin to 15cm long. Prophyll and peduncular bract are green with white scales. The peduncle is short, broadly elliptical in cross-section. Triads are borne almost throughout the rachilla, borne spirally at the base, becoming subdistichous toward the apex. Male flowers are light green in bud, glabrous, to about 9mm long, widely spreading at anthesis, stamens around 30, anthers white, pistillode longer than the stamens. Female flowers about 8mm wide. The fruit is ovoid, ellipsoidal to 2.2cm long by 8mm in diameter, orange to red at maturity, perianth persistent, stigmatic remains apical. The epicarp is smooth, mesocarp with longitudinal fibres, the endocarp is thin, the seed subglobose to ellipsoidal, raphe branches not prominent, the embryo basal.

Figure 33: Upper left. *Veitchia vitiensis* remains standing in forest cleared for pastures, Tholo-I-Suva, Viti Levu, Fiji.
Upper right. Leaf of *Veitchia vitiensis;* note the broad leaflets which are widely spaced.
Lower. *Veitchia vitiensis* in flower.

Veitchia arecina Becc. 1921

Locale, Habitat and Ecology:
Vanuatu, very commonly on the island of Nguna, less commonly on extreme northern Efate and Tanna, reported to occur on Pentecote, Erromango and Vanua Lava, from sea level to 500m in light moist forest on volcanic soils, occasionally overlying coral substratum.

Description: A medium palm to 18m with a grey trunk to 15cm in diameter, expanded at the base with a mound of adventitious roots, the internodes are short, the nodes raised. Leaves number about 11, ascending to horizontal and arching, apex not pendulous. The crownshaft is elongate, narrowing toward the petioles, densely white scaled with darker scales toward and into the short petiole which extends into an arching rachis similarly scaled below, glabrous above. Leaflets, about 50 per side, are held horizontally to semi-pendulous, glabrous above and below, the apex obliquely truncate and praemorse, the upper margin longer, brown scales on the mid and marginal veins below. Leaflets from 1.6cm to 7cm wide and from 15cm to 1m long, basally reduced to lorae.

The infrafoliar inflorescence is whitish in bud and fruit, about 1m long and 90cm wide, thrice branched with flexuous rachillae becoming semi-pendulous in fruit. The prophyll and enclosed peduncular bract are grey. The peduncle is elongate and broad, elliptical in cross-section, the arching rachis and rachillae are glabrous and angled. Triads are borne almost throughout, solitary males distal. Male flowers are oblong in bud, 1.5cm long by 7mm wide, the sepals about as long as the petals, stamens numerous. Fruit is ovate, bright red at maturity, 4cm long by 2cm in diameter, narrowed toward the apical stigmatic remains, perianth persistent. The epicarp is smooth, the mesocarp about 5mm thick, fibrous, the endocarp thin. The seed is subglobose, 2.5cm long by 1.8cm in diameter.

Veitchia pedionoma (A.C. Smith) H.E. Moore 1957
syn. *Vitiphoenix pedionoma*

Locale, Habitat and Ecology:
Fiji, endemic to Vanua Levu, in rainforest at 100-500m, in the vicinity of Latiki near Mt. Soro Levu and near the Korovuli River on the Seanggangga Plateau on the drier side of the island. It is an understorey to semi-emergent species.

Description:
A medium palm to 10m with a trunk diameter of 20cm. The crownshaft is elongate, the petiole to 50cm and the rachis to 3.5m are pale lepidote. Leaflets, about 50 per side, are crowded toward the apex, obliquely truncate and praemorse, glabrous above and below, with ramenta on the mid rib and marginal veins below, from 3.5cm to 5.5cm wide and 20cm to 75cm long, the basal leaflets reduced to lorae.

The infrafoliar inflorescence is glabrous, to 60cm long and broad, the rachillae ultimately to 15cm long, are thin and flexuous. Branches and rachillae are angled. Triads are borne almost throughout, spirally arranged basally, subdistichous toward the apex. Male flowers are white, to 8mm high, sepals much shorter than the petals, stamens around 35, the anthers about 4mm long. Female flowers about 3mm wide in bud with conspicuous bracteoles. The fruit is ellipsoidal, tapered to the base, bright orange to red at maturity, about 1.4cm long by 7mm in diameter, with persistent perianth, the epicarp is smooth, mesocarp is thickly fibrous and the seed is oblong.

Figure 34: *Veitchia arecina* on Nguna Island off the northern coast of Efate, Vanuatu (PC).

Figure 35: Following page. Upper left. *Veitchia simulans* in very wet forest on the central western slopes at about 300m as on Taveuni, Fiji
Upper right. Infructescence with mature fruit of *Veitchia simulans*.
Lower. Fruit of *Veitchia petiolata* collected from very wet forest northern Taveuni, Fiji.

Figure 36: Opposite following page. The crown of *Balaka seemannii.* note the broad cuneate leaflets and long peduncle on the inflorescence.

Veitchia petiolata (Burret) H.E. Moore 1957
syn. *Vitiphoenix petiolata*

Locale, Habitat and Ecology:
Fiji, on Vanua Levu, rarely on the southern slopes of Mt.Seatura in Mbua Province in dense forest at about 500m and on Taveuni in the north in dense very moist forest on rocky volcanic slopes at 300m.

Description:
A tall palm to 35m and trunk 20cm in diameter with an enlarged base. Leaves number about 7, are ascending to pendulous. The crownshaft is densely white scaled. The petiole, to 50cm long and the rachis to 2.25m long are similarly scaled below, the latter becoming pendulous, distally recurved. Leaflets, about 40 per side are closely spaced, pendulous, tapered from the middle to the base and the apex, obliquely truncate and praemorse, glabrous above and below with ramenta on the mid rib below and scales on the thickened marginal ribs. Leaflets are to 4cm wide and to 55cm long.

The infrafoliar inflorescence is brown lepidote, thrice branched with ultimate rachillae to 13cm, straight and slightly flexuous, the branches and rachillae angled. Triads are borne almost throughout, a loose spiral basally, distichous toward the apex. Male flowers are creamy white, about 7mm long, sepals much smaller than the petals, stamens around 50, the pistillode bottle shaped, attenuated, exceeding the stamens in length. Female flowers are subtended by conspicuous bracteoles. The fruit is orange to red at maturity, oblong to ellipsoidal, tapered to the base, to 2.2cm long by 1.1cm in diameter with a persistent perianth. The epicarp is smooth, the mesocarp fibrous and the endocarp thin and not attached to the seed which has a compressed base, a longitudinal raised hilum and a basal embryo.

Veitchia sessilifolia (Burret) H.E. Moore 1957
syn. *Vitiphoenix sessilifolia*

Locale, Habitat and Ecology:
Fiji, where it is endemic to Vanua Levu in the valley of the Yanawei River in Thakaundrove Province and near Navonu on the Natewa Peninsular as a semi- emergent in moist dense forest at 120-430m.

Description:
A medium palm to 10m with a grey trunk to 11cm in diameter, with internodes elongate, grey/green on the newly exposed upper trunk, the nodes raised. Leaves number about 6, ascending to horizontal and recurved. The crownshaft is narrow, to 50cm long, deep green, dense white lepidote with scattered black scales, denser toward the petiole which is to 26cm long, narrowed at the base and extending into a recurved rachis to 3.3m long, both are densely lepidote becoming glabrous with age. The leaflets, about 45 per side, are coriaceous, widely spaced, tapering from the middle to the base but less so to the apex which is broadly truncate, praemorse. The mid rib and marginal ribs are prominent on both surfaces, scales densely present on the mid rib and lepidote scales on the marginal veins. Leaflets measure from 1cm to 6cm wide and from 32cm to 1m long, the basal leaflets reduced to lorae.

The infrafoliar inflorescence, to 1.1m long, is 4 times branched, densely brown lepidote, ultimate rachilla straight and rigid or slightly flexuous to 16cm long. The peduncle is narrow, to 26cm long with parallel sides, elliptical in cross section, branches widely spaced, the rachillae closely grouped toward the apex. Triads are borne almost throughout, becoming distichous toward the apex, bracts inconspicuous. Male flowers are creamy-white to 9mm high, the sepals much shorter than the petals, stamens around 45, the anthers 5mm long. Female flowers, about 6mm high and 4mm wide, are subtended by prominent bracteoles. The fruit is oblong-ellipsoidal, to 2.2cm long and 1.2cm in diameter, narrowed toward the base, the apex with a cap of stigmatic remains. The perianth is persistent. The epicarp is smooth, the mesocarp thickly fibrous to 1.5mm thick, the endocarp thin and fragile. The seed

Figure 37: Silhouette of *Veitchia joannis,* Suva, Fiji.

is ovoid, apically pointed to 1cm long and 7mm in diameter, the hilum longitudinal and raised, raphe branches descending in parallel from the apex. The embryo is basal.

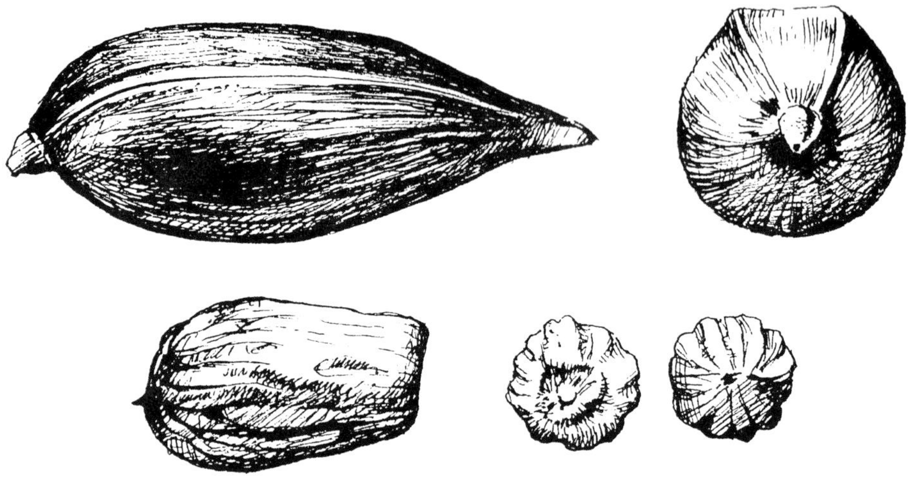

Veitchia sessilifolia. Endocarp (two views) and seed (two views)(three times natural size)

Veitchia joannis H. Wendl. 1868
syn.*Kentia joannis*
Kentia fipan

Locale, Habitat and Ecology:
Fiji, on the islands of Viti Levu, Vanua Levu, Taveuni, Matuku and Vanua Mbalavu where it is widespread and common as an emergent species forming dense or open colonies or as scattered individuals in dense or open forest from sea- level to 650m. This species reaches its culmination on ridges, crests and adjacent slopes in high rainfall areas on volcanic soils.

Description:
A tall palm to 32m with a grey/brown trunk to 28cm in diameter with an expanded base, densely scaled on the upper internodes. Leaves number about 9, ascending, horizontal or arching, with a recurved apex. The crownshaft is elongate to 1.2m long, narrowing toward the petioles, green with grey tomentum, dark scales becoming denser toward the petiole, particularly near the margin opposite the petiole. The petiole, to 25cm long, usually much less, is densely pale brown scaled, those below softer and interspersed with large dark brown scales. The rachis, to 3.3m long, is similar but becomes waxy toward the apex. Leaflets, about 100 per side, are closely spaced, arching pendulous to pendulous, tapering from the middle to the base and the apex which is obliquely truncate and praemorse, glabrous above and below with brown ramenta and small scales on the mid rib near the base. Leaflets measure from 24cm to 93cm long by 8mm to 8cm wide, the basal leaflets reduced to lorae. The prophyll has an ear-like projection at the base, is flat, with dense white floccose tomentum, the peduncular bract enclosed within, similarly shaped and surfaced.

The erect infrafoliar inflorescence to 60cm long, is green with brown tomentum persisting at anthesis and in fruit, the rachillae flexuous, ultimately to 20cm long, narrowing abruptly at the apex. The peduncle, to about 10cm long is narrowly elliptical in cross-section, the branches and rachillae angled, becoming semi- glabrous in fruit. Triads are borne on the few basal nodes, paired or solitary males distally, bracts inconspicuous. Male flowers are greenish/ivory in bud to 1.7cm high, widely spreading and white at anthesis, the sepals much shorter than the petals, stamens around 100, the pistillode bottle shaped, trifid at the apex, exceeding the stamens in length. The females are about 1.5cm high and 1cm in diameter. The fruit, red to crimson at maturity, is about 6 cm long and 3cm in diameter, rounded at the base, attenuated apically to a distinct pointed cap with stigmatic remains. the flaring perianth is persistent. The epicarp is smooth, the mesocarp strongly fibrous, about 2mm thick, the endocarp is thin and fragile, adherent to the seed. The seed is ovoid to 3.6cm long by 1.9cm in diameter, pointed at the apex with a longitudinal raised hilum, the raphe also longitudinal with branches descending from the apex. The embryo is basal.

Veitchia metiti Becc. 1921

Locale, Habitat and Ecology:
Vanuatu, where it is presumed endemic to Vanua Lava at 915m, in very moist forests on volcanic soils. This species is incompletely known and the description below is based on fragmentary material.

Description:
The fragment of inflorescence is stout, flattened, to 1.2cm wide, 6mm thick at the base, with flattened and angular rachillae to 10cm long, 4mm thick, triads borne on the lower nodes, the upper part flexuous with male flowers only. Fruit is ellipsoidal, 4.5cm long by 2cm in diameter, tapered to the base and the apex with stigmatic remains on a cap to 3mm long, with a persistent perianth. The seed is rounded at the base and the apex, 2.3cm long by 1.6cm wide, the raphe branches descending and well branched.

BALAKA Becc. 1885

A genus of about seven species, the majority in Fiji, the remainder in Samoa. Five species are annotated in this work, one of which is not well known botanically. This genus is closely related to *Ptychosperma* and *Drymophloeus. Balaka seemannii* is the type species.

Key to the Species

Leaflets narrowing from the middle to the apex
- Fruits less than 1.8cm long
  - Rachillae less than 9cm long with triads borne basally
    Ovalau — *B.pauciflora*
  - Rachillae greater than 10cm long with triads throughout
    Viti Levu — *B.microcarpa*
- Fruit greater than 2cm long
  Vanua Levu, Viti Levu — *B.macrocarpa*

Leaflets broad at the apex
- Endocarp keeled
  Viti Levu — *B.longirostris*
- Endocarp not keeled
  Vanua Levu, Taveuni — *B.seemannii*

Balaka pauciflora (H.Wendl.) H.E. Moore 1957

syn. *Ptychosperma pauciflorum*
Vitiphoenix pauciflora
Drymophloeus pauciflorus

Locale, Habitat and Ecology:
An incompletely studied species reported to occur on Ovalau, Fiji, in moist forest as an understorey palm.

Description:
With a trunk to 2.5cm in diameter, leaflets broadly linear, narrowing toward both the base and the apex which is toothed.

The inflorescence has rachillae to 9cm long with triads borne only on the basal two-fifths, female flowers to 6mm high. The fruit is 1.5cm long by 9mm in diameter.

Notes:
Although herbarium material of this species has not been located and confusion surrounds the original notes and citations, this species still remains valid.

Balaka microcarpa Burret 1940
syn. *Balaka microcarpa* var. *longicupsis*
Balaka microcarpa var. *microcarpa*

Locale, Habitat and Ecology:
Fiji, Viti Levu, where it is a scattered but common species in the very moist rainforests of the south-eastern hills at about 100-200m. It is an understorey to semi-emergent component, growing with *Veitchia vitiensis* and *Balaka longirostris.*

Description:
An occasionally tall palm to 13m with a characteristically white-moss covered trunk to about 8cm in diameter. The leaves are ascending, arching to 1.5m long. The crownshaft is dark green, mottled in lighter green and slightly bulging. The petiole and rachis are covered in greyish tomentum, which is deciduous. Leaflets are elongate sigmoidal, narrowing from the middle to both the base and the apex, which is toothed. Leaflets are rigid with a very prominent mid rib and measure to 56cm long by 10.5cm at the middle.

The infrafoliar inflorescence, with an elongate peduncle, held obliquely erect, is branched 2-3 times with flexuous rachillae to 23cm long. Triads are borne throughout, flowers lepidote, the males being red/brown to 5mm long with a long canary yellow pistillode and with 28-32 stamens, the females to 5mm high. Flowers are distichously arranged. Fruit is irregularly ovoid, attenuated toward the apex, 1.6cm long by 8mm wide, the seed with minor ridges and elongate hilum.

Balaka macrocarpa Burret 1935

Locale, Habitat and Ecology:
Fiji, on Viti Levu and Vanua Levu, occurring uncommonly in dense moist forests to about 800m in areas of high rainfall in the mountainous interiors as an understorey palm. On Viti Levu, it grows with *Cyphosperma tanga* in the vicinity of Mt.Lomalangi and Tomanivi in the north of the island at 850m.

Description: A smallish palm with a trunk to 8m and a diameter of 6cm. The leaves are ascending to about 2m long. The crownshaft is dark green, mottled in lighter green and slightly bulging. The petiole, 6-24cm long, and the rachis are covered with tomentum which is deciduous. Leaflets are rigid, held erectly to horizontally, elongate-sigmoidal, narrowing from the middle to both the base and the apex which is toothed, measuring to 39cm long and 6cm wide, with a prominent mid rib.

The infrafoliar inflorescence, with an elongate peduncle, held obliquely erect has flexuous rachillae to 33cm long. Triads are borne throughout, male flowers lepidote with numerous stamens, the females with petals to 1.2cm high. The fruit is orange, irregularly ovoid, to 4.2cm long and 1.7cm in diameter.

Balaka longirostris Becc. 1914
syn. *Balaka leprosa*

Locale, Habitat and Ecology:
Fiji, on Viti Levu in rainforest from near sea-level to over 1,000m as an understorey palm, most abundant in the mountainous interior particularly at higher altitudes around Nadarivatu and Navai, uncommonly at lower altitudes near Ngaloa in the south and more uncommonly in the hills and mountains north of Suva. This species prefers moist well drained soils in sheltered sites in gullies and slopes near streams.

Figure 38: *Balaka microcarpa* in residual forest at Tholo-I-Suva, south Viti Levu, Fiji.

Description:

A smallish wispy palm with a trunk to 7m, green with prominent raised nodes of irregular width and disposition, and to a diameter of 5cm. Leaves number about 7, are ascending to horizontally held. The crownshaft is narrow and modest, greyish/green, with deciduous tomentum, the petiole to 26cm long and the rachis to 1.8m long, similarly scaled, more so beneath. Leaflets are cuneate, apically toothed, with curved margins, dark green above and grey scaled beneath, mid rib is prominent and lateral veins raised on the upper surface. The apical leaflets are united, the mid leaflets measuring to 27cm long and to 16cm wide.

The infrafoliar inflorescence has an elongate peduncle, 1 to 2 times branched with flexuous rachillae to 22.5cm long. The prophyll and peduncular bract are papery, the latter attached toward the apex of the former, both bracts deciduous early. Branching of the inflorescence is more or less alternate and distichous, the rachilla having a bare basal portion. Triads are borne to the apex, positioned distichously, male flowers to 7mm long are widely opening at anthesis, with a long wiry pale yellow pistillode and from 25-44 stamens, female flowers with petals to 1.4cm long. The fruit are fusiform-ellipsoidal with an attenuated apex and apical stigmatic remains, clear red carmine at maturity, to 3.8cm long and 1.6cm in diameter, crowded distichously on the rachillae. The endocarp is thin. The seed, to 3cm long, has prominent longitudinal ridges and a rostrum projecting from the apex.

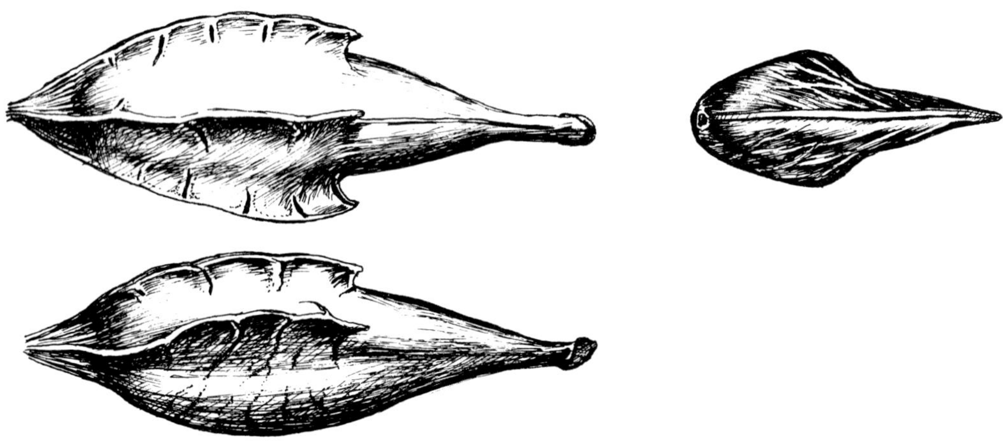

Balaka longirostris. Endocarp (two views) and seed (three times natural size)

Balaka seemannii (H.Wendl.) Becc. 1885

syn. *Ptychosperma seemanni*
Ptychosperma perbreve
Balaka perbrevis
Kentia kersteniana
Balaka gracilis
Drymophloeus seemannii
Vitiphoenix seemanni
Balaka cuneata

Locale, Habitat and Ecology:

Fiji, on Vanua Levu and Taveuni, where it occurs uncommonly as scattered individuals or small groups as an understorey palm in very wet forests in dense shade from sea level to about 1,000m. On Taveuni it is most common on the eastern slopes of the mountainous backbone where the annual rainfall exceeds 7,000mm. *Veitchia petiolata* occurs in the vicinity.

Figure 39: Upper left. Juvenile of *Balaka microcarpa.*
Upper right. Crown of *Balaka longirostris,* cultivated in the garden of R.H. Phillips, Suva, Fiji.
Lower. Inflorescence of *Balaka longirostris* with male flowers at anthesis.

Description: A smallish palm with a dark green trunk to about 8m, the leafscars prominently raised, evenly disposed, and a diameter to 5cm. Greyish lepidote scales and scattered green scales persist on the internode beneath the fallen leaf for only a short time. The leaves number around 7, ascending to horizontal, to about 2m long. The crownshaft is narrow and modest, densely scaled with dark green lepides, more so in the upper portion and into the petiole, to 25cm long and less so on the rachis to about 1.75m long. The leaflets number to about 12 per side, are cuneate, apically toothed, dark green above, with a prominent mid rib and thick marginal veins, measuring to 23cm in length to 10cm wide at the apex. The apical leaflets are united.

The infrafoliar inflorescence has an elongate peduncle, 1 to 2 times branched with flexuous rachillae to 23cm long. The prophyll and peduncular bract are papery, the latter attached toward the apex of the former, the prophyll occasionally persistent. Both bracts are covered in greyish/green tomentum. Triads are borne to the apex, a bare basal portion on once branched branches, distichously arranged. Male flowers, to 6mm long, are glabrous to lepidote, widely spreading at anthesis, a long whitish flexuous pistillode and 21-29 bright yellow stamens. Females have petals to 6mm high. The fruit is curved-ellipsoidal, with an attenuated apex and apical stigmatic remains, bright red at maturity, distichously crowded on the rachilla. The endocarp is thin and ridged. The seed is similarly shaped, with longitudinal ridges but lacking a projecting keel.

Balaka seemannii. Endocarp (two views)(twice natural size)

Figure 40: Upper left. Leaf of *Balaka seemannii* from northern slopes of Taveuni, Fiji.
Upper right. Maturing fruit of *Balaka seemannii.*
Lower. Male flowers of *Balaka seemannii.*

Figure 41: Following page. *Gulubia cylindrocarpa* at about 200m asl on Mt. Garigona, Vanua Lava, Vanuatu.

Figure 42: Opposite following page. Upper. The leaves of *Gulubia cylindrocarpa* remain entire or irregularly divided until the palm reaches and emerges from the canopy.
Lower left. Inflorescence of *Gulubia cylindrocarpa* with male flowers at anthesis; note the long petals.
Lower right. Mature fruit of *Gulubia cylindrocarpa.*

ARECINAE

Subfamily: Arecoideae
Tribe: Areceae

Loxococcus
*Gulubia**
Gronophyllum
Nenga
Siphokentia
Pinanga
Hydriastele
Areca
* Indigenous genus

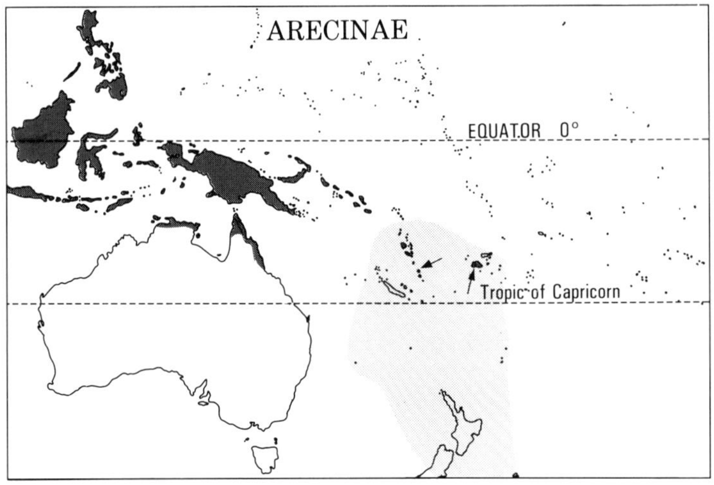

Description:
Diminutive to very tall, understorey to emergent; solitary or clump-forming palms; leaf sheathing, simple pinnately veined or pinnately leaved; unarmed; monoecious, pleonanthic; rachillae straight, not widely spreading; endocarp lacking an operculum.

Distribution:
From equatorial to tropical; sea-level to moderately high altitudes; rainforest, moist forest, swamp forest, stream edges, monsoon forest, coastal forests. Sri Lanka, north-east India, South-East Asia, Malesia, Philippines, New Guinea, tropical Australia, Solomons, Vanuatu and Fiji.

Notes: This SUBTRIBE of eight genera is very well developed in Malesia and New Guinea with only small or monotypic genera found outside of the centre of diversification. *Pinanga,* centred in Malaysia, has at least 105 species. There is diversity displayed with leaf anatomy. Some genera e.g. *Pinanga, Areca* and *Gronophyllum,* have species with simple bifid leaves or largish, by comparison, leaves with many leaflets. *Gulubia* is unusual in that some species have lax pendulous leaflets, not to be found in the South-West Pacific, where others have erect rigid leaflets.

GULUBIA Becc. 1885

A genus of about nine species, distributed through east Malesia, New Guinea, Australia, Solomon Islands, Palau Island, Vanuatu and Fiji.The closest relatives are *Hydriastele* and *Gronophyllum,* the former distinct because of its clustering habit, although the flowers are essentially similar, the latter differing in the flowering mode, being protogynous, i.e. female flowers achieve receptivity prior to male anthesis

Key to the Species

Fruit greater than 1 cm long, pericarp thickish
    Fruit yellowish at maturity
        Vanuatu, Santa Cruz Group — *G. cylindrocarpa*

Fruit less than 1 cm long, pericarp thin
    Fruit whitish at maturity
        Fiji — *G. microcarpa*

Gulubia cylindrocarpa Becc. 1910

Locale, Habitat and Ecology:
Vanuatu, commonly in dense colonies on Ureparapara, commonly as scattered individuals and small groups on Vanua Lava, Malekula and Erromango as well as the Santa Cruz Group of the Solomon Islands. It is an emergent component of ridges and crests in rainforest and *Agathis* forest on volcanic soils at 50-800m. Juveniles develop stilt roots and undivided leaves, both of which diminish in formation with maturity.

Description:
A tall palm to 27m with a light grey trunk with closely spaced leafscars, and 25cm in diameter and broadened base. The arcuate leaves, to about 3m long, number to 18. The crownshaft, with reddish/white scales most dense in the upper portion, is glaucous, light green to 95cm long. The petiole, to 25cm, and the arching rachis, to 2.75m, have similar scales. The erect, apically notched leaflets number to 55 per side, are light green above and grey-green beneath, with whitish ramenta on the base of the mid rib below, measure to 1m long and 3.3cm wide, with a prominent mid rib. The leaflets in the apical one-third are laterally twisted.

The infrafoliar inflorescence has up to 32 almost parallel rachillae, ultimately to 43cm long and 2.5cm in diameter. The broadly based peduncle is winged, constricting above the wings to a narrow rachis with close bract scars. The inflorescence is at first erect but becomes pendulous in fruit. Triads are borne in vertical rows to the apex of the rachilla excepting a bare basal portion. The males, not widely opening, have soft pointed petals to 2cm long, the outside one slightly longer, lacking a pistillode and with 6 stamens. The female flowers are smaller, to 4mm high and 2.25mm wide. The curved cylindrical fruits, dull yellow at maturity, are to 1.5cm long and 6mm in diameter, borne in vertical rows, with apical stigmatic remains and longitudinal fibres in the mesocarp. The seed is similarly shaped, to 1cm long and 4mm in diameter with an elongate hilum.

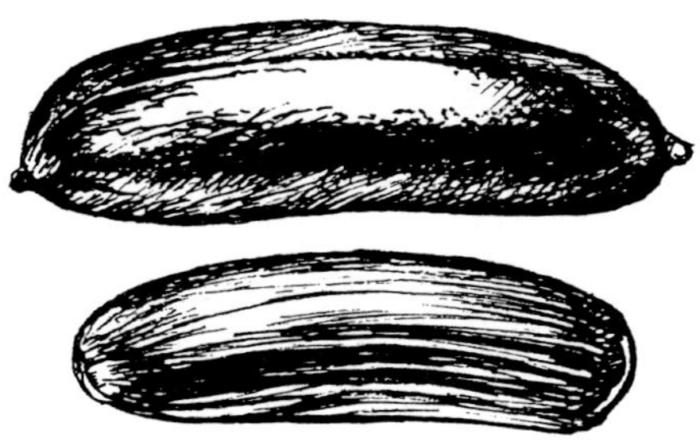

Gulubia cylindrocarpa. Fruit and endocarp (four times natural size)

Gulubia microcarpa Essig 1982

Locale, Habitat and Ecology:
Fiji, on the island of Viti Levu, confined to loose colonies on ridges and slopes inland from Ngaloa in the south, as an emergent species in moist forest at 230- 300m.

Description: A tall palm with a grey trunk to 25m and a diameter of 28cm. The arcuate leaves number about 17. The crownshaft, to about 80cm long, is light green, the petiole to 30cm and the rachis to 2.2m are brown lepidote with thin whitish tomentum. The leaflets are erect, about 52 per side and measuring to 1.1m long and 3.5cm wide, with a bifid/praemorse apex and ramenta on the mid rib below.

The inflorescence is held horizontal, well branched with rachillae to 42cm long and 2.5cm in diameter. Males flowers not collected. Female flowers are 2mm high. The fruits measure 1cm long by 3mm in diameter, cylindrical with a slight curve and whitish at maturity. The mesocarp is thin with flattened fibres, the seed to 8mm long by 2mm wide with an elongate hilum.

Notes: Male flowers of this species have yet to be collected. It is closely related to *G. cylindrocarpa* of Vanuatu and the Santa Cruz Group.

IGUANURINAE
J.D. Hooker

Subfamily: Arecoideae
Tribe: Areceae

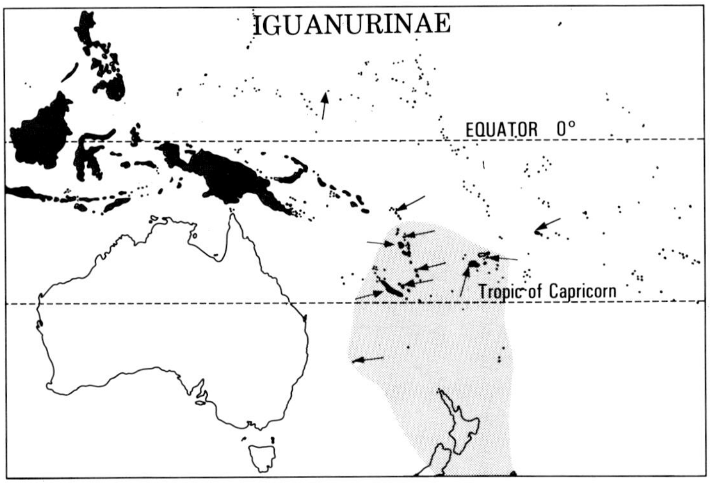

*Neoveitchia**
*Moratia**
*Campecarpus**
*Pelagodoxa**
*Clinostigma**
*Basselinia**
Iguanura
*Alsmithia**
*Cyphosperma**
*Brongniartikentia**
Satakentia
*Veillonia**
*Lepidorrachis**
*Rhopaloblaste**
*Burretiokentia**
*Heterospathe**
*Dictyosperma**
*Physokentia**
Sommieria
Actinorhytis
*Goniocladus**
Bentinckia
*Lavoixia**
*Carpoxylon**
*Clinosperma**
*Alloschmidia**
*Cyphokentia**
*Cyphophoenix**
* Indigenous genera

Description:
Small to very tall, understorey to emergent, solitary or clump forming palms; leaf non-sheathing, partly sheathing or sheathing, simple pinnately veined or pinnately leaved; unarmed; monoecious, pleonanthic; flowers unisexual; endocarp possessing an operculum.

Distribution:
From equatorial to warm temperate; sea-level to moderately high altitudes; rainforest, moist forests, swamp forests, coastal forests. Mascarene Islands, South India, Andaman and Nicobar Islands, South-East Asia, Malesia, Philippines, Ryukyu Islands, New Guinea, Solomons, Caroline Islands, Vanuatu, New Caledonia, Lord Howe Island, Fiji and Marquesas Islands.

Notes:
This SUBTRIBE is characterised by having a large proportion of monotypic or limited genera which are disjunctly distributed on sometimes very isolated oceanic islands. Of the 28 genera, 20 are found in the South-West Pacific, 17 of which are endemic, *Pelagodoxa, Clinostigma* and *Physokentia* also occuring in nearby areas. It may be assumed on the basis of distribution that the South-West Pacific is the centre of development of this SUBTRIBE.

NEOVEITCHIA Becc. 1920

A monotypic genus endemic to Fiji, with obscure affinities, but displaying similarities to *Actinorhytis* and *Rhopaloblaste* of New Guinea and the Solomon Islands, *Carpoxylon* of Vanuatu, *Lepidorrachis* of Lord Howe Island and *Clinostigma,* widespread throughout the west and south-west Pacific. It holds an isolated position within Iguanurinae.

Neoveitchia storckii (H.Wendl.) Becc. 1920

syn. *Veitchia storckii*

Locale, Habitat and Ecology:
Fiji, on Viti Levu, where it is restricted to a small area of about 10sq km in the Rewa River Valley to the north of the village of Viria in Naitasiri Province, an emergent palm on the slopes and in the gullies in moist forest. Listed as endangered by the I.U.C.N., parts of its range have recently (1987) come under the jurisdiction of Fijian Crown Land, thus ensuring conservation. It has been successfully introduced into cultivation.

Description:
A largish palm with a closely scared, light brown becoming grey trunk to 12m, often covered in grey moss, to a diameter of 25cm and expanded to 45cm at the base. Leafscars are irregularly disposed, undulating, but not prominent. Leaves, to 5m long, number 12-15, erect to horizontal with a lateral twist to almost 90°. Crownshaft is incompletely developed, glossy dark green to black, split opposite the petiole, the margins becoming fibrous with age. The acuminate leaflets number about 70 per side, to 75cm long and 8cm wide, glossy green above, brown lepidote beneath, with a prominent yellowish mid rib, thickened margins and numerous veins of lesser prominence. Reddish/brown ramenta is present on the mid rib below.

The infrafoliar inflorescence spreads to 80cm, is 2 to 3 times branched, white in bud and floral anthesis, becoming brown/green in fruit. The prophyll, glossy dark green, held erect, encloses a thin, brown lepidote rostrate peduncular bract, exserted by about 25cm, together longer than the crownshaft and extending beyond the petioles. The peduncle, horizontally held, to about 13cm long, is broad based, elliptical in cross-section, constricting beyond the bract scars into a short rounded rachis to 20cm, the ultimate rachillae flexuous and pendulous to 45cm long. Triads are borne on the basal 7cm only, the larger females positioned below the males, paired males in vertical rows for about 32cm, deciduous in fruit. Rachillae and triads are subtended by acute bracts, those of the male rounded. Male flowers, to 3mm wide in bud, 4.5mm at anthesis, have imbricate sepals, valvate petals and a pistillode about length of the stamens of which there are 6. Female flowers are to 8mm high and wide. Fruit is ellipsoidal, to 5cm long and 2.2cm in diameter, attenuated to the apex with apical stigmatic remains, becoming reddish/yellow at maturity, mesocarp has thick longitudinal fibres attached to the thin woody endocarp. The operculum is basal, small and circular. The seed, with an elongate hilum, is to 2.5cm long by 1.5cm in diameter, with a basal embryo.

PELAGODOXA Becc. 1917

A supposedly monotypic genus native to the Marquesas Islands where it is rare. The inclusion in this work is of a single occurrence in Vanuatu of a taxon which may prove, with further study and collection, to be conspecific with *P. henryana,* despite difference in inflorescence and fruit dimensions.

Pelagodoxa sp. aff. ***henryana***

Locale, Habitat and Ecology:
Vanuatu, on Vanua Lava in the Banks Group, in dense secondary forest at about 250m in the vicinity of an abandoned village. This palm was collected and described (see below) by Dr. J.-M. Veillon of ORSTOM, Noumea on 14/6/1983.

Description:
J.-M. Veillon 5567. Two sheets in the Tagabe Herbarium, Port Vila, Vanuatu; the following notes by the collector.

Single stemmed erect palm. Trunk to about 8m high and 15cm in diameter, slightly conic expanded at base (to about 35cm at ground level) with some prominent roots. Leaves 27 on the tree. 2.5-3m long; 60-70 pairs of leaflets, entirely united; blade grey-glaucous beneath, shining dark green upside - 2 axillary inflorescences 50-60 by 30-40cm. Fruit not seen, but about 2cm in diameter when ripe (from indigenous people).

BRONGNIARTIKENTIA Becc. 1920

A genus of two species endemic to New Caledonia, with obscure affinities but sharing some characteristics with *Clinosperma* and *Cyphosperma. Brongniartikentia vaginata* is the type species.

Key to the Species

Inflorescence pendulous with an elongate peduncle
 Floral bracts with ciliate margins
  Prophyll and peduncular bract persistent
   Petiole short
    B. vaginata

Inflorescence erect with peduncle and rachis equal
 Floral bracts smooth
  Prophyll and peduncular bract early deciduous
   Petiole long
    B. lanuginosa

Brongniartikentia vaginata (Brongn.) Becc. 1920
syn. *Cyphokentia vaginata*
Clinostigma vaginata

Locale, Habitat and Ecology:
New Caledonia on La Grande-Terre, in the southern sector, most commonly in the areas to the east of Riviere Bleue, but generally uncommon in other areas, as an understory palm in moist forests on ultrabasic, peridotitic soils at from 200- 1,000m.

Description:
A smallish palm with a slender, prominently noded, green trunk to 5m and 5cm in diameter. The remains of old inflorescence bases are persistent. The arching leaves number about 6, the leafbases forming an incomplete crownshaft split opposite the petiole, being to 55cm long, deep green and loosely scaled. An ear- like growth is present near the petiole, which is densely scaled to about 25cm long, the rachis to 1.8m similarly scaled. The coriaceous leaflets number about 16 per side, regularly but distantly arranged at intervals to 10cm, and measuring from 7mm to 5cm wide by 9cm to 75cm long.

The arching pendulous inflorescence, to 1.5m long, maturing whilst the leaf, in the axil of which it is inserted, remains attached, is persistent and the base reduced to a fibrous stub with age. The prophyll, to 35cm long and 5cm wide, is lepidote at the apex. The peduncular bract is inserted about 20cm above the prophyll insertion and exceeds it by up to 55cm. Both are persistent. The peduncle is reddish/brown, floccose becoming glabrous and the rachillae, to 50cm long, have triads which are subtended by ciliate margined bracts to about the length and paired males distally. Male flowers, to 1.5mm high with reddish valvate petals, have 6 stamens and an apically broad pistillode which is longer than the stamens. Female flowers are slightly larger. The fruit is ovoid- ellipsoidal, to 1.9cm long by 8mm wide, with stigmatic remains being basal. The endocarp is keeled at the base near the circular operculum. The seed is similarly shaped, but without the keel, to 1.1cm long by 6mm in diameter, the embryo basal.

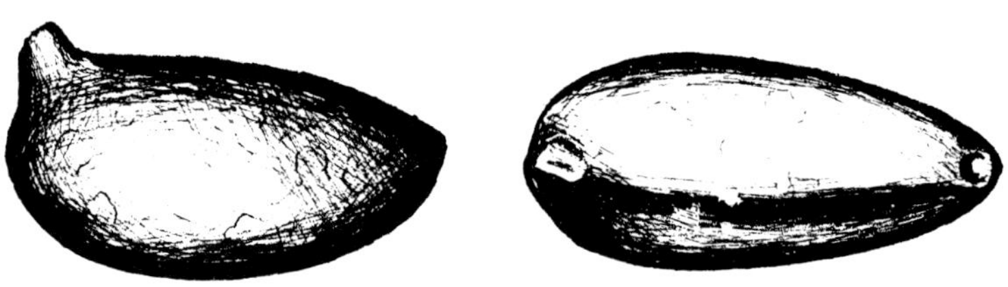

Brongniartikentia vaginata. Endocarp (two views)(four times natural size)

Figure 43: *Neoveitchia storckii* in flower and with young fruit, cultivated in the grounds of FINTEL, Suva Point, Viti Levu, Fiji.

Brongniartikentia lanuginosa H.E. Moore 1974

Locale, Habitat and Ecology:
New Caledonia, in the north-east sector, on Mt.Panie, Mt.Ignambi and Mt.Colnett and nearby areas, as an understory or semi-emergent palm at 700-1,250m in moist forests on schists or siliceous soils. On Mt.Panie it grows with *Moratia cerifera* and *Chambeyronia lepidota.*

Description:
A medium palm with a ringed trunk to 6m and 10cm in diameter. The spreading arching leaves number about 5, the leafbases not forming a distinct crownshaft, but split opposite the petiole. The leafbase is densely tomentose with brown scales. The petiole is long, to 60cm and the arching rachis to 1.5m is similarly covered as the leafbase. The coriaceous, glossy, dark green leaflets number about 20 per side, acute with twisted ramenta on the prominent mid rib beneath.

The erect inflorescence emerges from within the leaves, usually maturing whilst the leaf, in the axil of which it is inserted, remains attached to the trunk, but also after the leaf has fallen. The prophyll and the peduncular bract are early deciduous, tomentose with brown scales, the former to 27cm long, splitting at the base, and the latter exceeding the prophyll to about 28cm long. The glabrous peduncle, branches and rachillae, ultimately to 26cm long, are all erectly ascending and remain so in fruit. Triads are borne on the lower two-thirds of the rachillae, paired or solitary male flowers distally. Liplike bracts to 2mm long subtend the triads. Male flowers are about 2mm high, with 6 stamens, the females red in bud to about 3mm high. The ovoid-ellipsoidal fruit measure to 2.3cm long and 1.2cm in diameter, with stigmatic remains basal. The endocarp is basally keeled, the operculum circular. The seed is similar but without the keel, and a basal embryo.

Brongniartikentia lanuginosa. Fruit and endocarp (natural size)

CLINOSPERMA Becc. 1920

A monotypic genus endemic to New Caledonia. It appears most closely related to *Basselinia,* also of New Caledonia and *Lepidorrachis* of Lord Howe Island.

Clinosperma bracteale (Brongn.) Becc 1920
syn. *Cyphokentia bractealis*
Clinostigma bracteale

Locale, Habitat and Ecology:
New Caledonia, endemic to the southern half of La Grande-Terre where it is widespread but uncommon as a semi-emergent species in moist to very moist forests on serpentine and schistose soils at altitudes from 40-1,200m.

Description: A tall palm to 17m with a green, prominently ringed trunk to 10cm in diameter. The spreading leaves number about 4 to 6 with a narrow not fully developed crownshaft, deeply split

Figure 44: *Brongniartikentia lanuginosa* at 700m asl on Mt. Panie', north-eastern New Caledonia (O).

Figure 45: Following page. The leafsheaths of *Neoveitchia storckii.*

Figure 46: Opposite following page. The crown of *Pelagodoxa* sp. aff. *henryana,* a solitary specimen growing at about 250m asl on slopes above Chelva River, Vanua Lava, Vanuatu (O).

opposite the petiole, to 70cm long, greyish/green with brown scales toward the base. The petiole, circular in cross-section, to 65cm long and the rachis to 1.5m, are reddish, covered with scales and tomentum. The leathery, acute leaflets number about 35 per side, are glossy green above and paler beneath with a prominent mid rib which lacks ramenta.

The infrafoliar inflorescence is compact, greenish/brown, well branched with ultimate rachillae to about 23cm long. The prophyll is prominently keeled, fully enclosing the peduncular bract. A second peduncular bract is often present. Branches and rachillae are subtended by acute bracts 1-10cm long. Triads are borne in the lower one-third of the rachillae, the small male flowers are dull red in bud to 1.5mm long with 6 stamens and a broad pistillode which is as long as the stamens. The female flowers, also red, are about twice the size of the males at anthesis. The fruits are globose, black at maturity, with a smooth epicarp and lateral stigmatic remains. They measure to 1.6cm in diameter. The endocarp has a basal operculum. The seed is globose, about 9mm in diameter, with a basal hilum and basal embryo.

Clinosperma bracteale. Fruit (natural size)

LEPIDORRACHIS (H.A. Wendl. and Drude) O.F. Cook 1927

A monotypic genus endemic to Lord Howe Island, having an affinity with *Clinostigma* and *Clinosperma.*

Lepidorrachis mooreana (F.v.Muel.) O.F. Cook 1927
syn. *Clinostigma mooreanum*
Kentia mooreana

Locale, Habitat and Ecology:
Lord Howe Island, where it is restricted to on and below the windswept and misty summits of Mts. Gower and Lidgbird above 750m. The palms are scattered throughout the subtropical cloud forest which is predominantly of smaller gnarled moss-covered trees such as *Bubbia howeana* and *Dracophyllum fitzgeraldii.* The soil is poorly developed, overlying deeply weathered basalt. The dense canopy formed by the 2 endemic trees mentioned is no more than 5m in height. Tree ferns (*Cyathea* spp.), terrestrial ferns, lichens and mosses are abundant with large tussock sedges being common. This palm emerges through the canopy on cliff edges and exposed sites where it receives full sunlight and strong buffeting winds.

Description:
A small palm with a green/grey closely ringed trunk to 3m and 15cm in diameter. The arching spreading leaves number about 8. The leafbases are broadly clasping, not forming a distinct crownshaft, deep green to pinkish, densely scaled toward the petiole which is short but broad, extending into an arching twisted rachis to 1.5m long, surfaces of both with brown tomentum. The coriaceous linear acute leaflets are held erectly, to 3cm wide and to 60cm long, with a prominent mid rib, thickened margins and 2-3 secondary veins, both surfaces of the mid rib scaled.

The infrafoliar inflorescence is densely scaled, basally branched, with short, rigid rachillae. Peduncle is short, rachis longer. The prophyll is green, keeled, to 30cm long, not fully enclosing the white to light green, beaked peduncular bract. Triads, subtended by acute bracts, are borne almost throughout the rachilla. The males cream, to 6mm high, with a cylindrical pistillode exceeding the stamens of which

Figure 47: *Clinosperma bracteale* in habitat, New Caledonia (LHBH).

there are 6. The females are larger than the males. Fruits are globose, to 1 cm in diameter, red at maturity. The endocarp is operculate, to 7mm in diameter with an elongate raphe and apical hilum. Embryo is sub-basal.

Lepidorrachis mooreana. Endocarp (two views)(six times natural size)

ALLOSCHMIDIA H.E. Moore 1978

A monotypic genus endemic to New Caledonia, with obscure affinities but some similarities with genera such as *Cyphophoenix* and *Campecarpus,* also of New Caledonia.

Alloschmidia glabrata (Becc.) H.E. Moore 1978
syn. *Basselinia glabrata*
Alloschmidia radicosa

Locale, Habitat and Ecology:
New Caledonia, where it occupies disjunct altitudes, at about sea-level at the foot of Mt. Panie and then at 240-1,170m in wet forests on schists and micaschists on Mt. Panie and other nearby mountains of the north-east sector of La Grande- Terre. This is the wettest region of New Caledonia, where palms are abundant, particularly at moderate to high altitudes. This is a semi-emergent species, growing in association with *Veillonia alba* and *Chambeyronia lepidota.*

Description: A tall palm to 15m with a lightly ringed trunk to 15cm in diameter, with prop roots at the base. The leaves, to 2.15m long, number about 10, ascending. The crownshaft is elongate to 90cm, constricted toward the petiole, green with green/brown scales. The petiole, to 15cm, and the rachis, to 2m, are similarly scaled. The acute linear leaflets number to about 25 per side, evenly distributed, with the basal few being reduced to narrow lorae. Leaflets are to 25cm to 1.05m long by 2.5cm to 7cm wide, lepidote on both surfaces, more so below, with ramenta on the mid rib beneath.

The infrafoliar inflorescences number up to 17, but usually about 12, the lower- most in mature fruit, the upper-most in floral bud. They are ranked either distichiously or spirally, more or less in alternate nodes. The peduncle, semi- circular in cross-section, to 25cm long, is held horizontally, glabrous with age. The pendulous rachillae, to 45cm long, are thin. The prophyll is tubular, yellowish green, enclosing a distally tomentose peduncular bract and one incomplete bract. Triads are borne in horizontal lines almost throughout, lacking bracts, both sexes being brown, the males to 2mm high, with an angular, cylindrical, pistillode slightly exceeding the stamens of which there are 6, and with females to 3mm high. The fruits, to 1.5cm long and 5mm in diameter are curved ellipsoidal with apical stigmatic remains, black on maturity with a non- adherent perianth. The endocarp is pitted, fragile, the seed is narrowly ovoid- ellipsoidal, with an elongate operculum, lateral embryo and elongate hilum.

Figure 48: Upper left, Infructescence of *Brongniartikentia vaginata,* growing in the Riviere Bleue basin, New Caledonia (AD)
Upper right. Fruits of *Brongniartikentia vaginata* (AD).
Lower. Crown of *Alloschmidia glabrata;* note the number of inflorescences and their ranked arrangement (O).

CLINOSTIGMA H.A. Wendl. 1862

A genus of about thirteen species distributed from the Caroline Islands, Bonin Island to eastern New Guinea, the Solomon Islands, Vanuatu, Fiji and Samoa. Two species to be annotated in this work, one from Fiji and one from Vanuatu.

Key to the Species

Inflorescence exceeds 80cm in length with 1st order branches orientated to 45° or more
Fruit black with subapical stigmatic remains
Vanuatu
C. harlandii

Inflorescence less than 60cm in length with more or less parallel 1st order branches
Fruit red with apical stigmatic remains
Fiji
C. exorrhizum

Clinostigma harlandii Becc.

Locale, Habitat and Ecology:
Vanuatu, on the islands of Vanua Lava, Espiritu Santo, Pentecote, Efate, Erromango and Aneityum (?) as an emergent palm in very moist forests on ridges, crests and steep slopes at from 400 to over 1,000m, but most common in the upper altitudes of its range.

Description: A tall palm with a green/grey trunk to 15m, to a diameter of 25cm, 35cm at the base, with short internodes, those below the crownshaft whitish. Leaves, numbering about 10, are arching to pendulous to 5m long. The crownshaft is narrow, glaucous/green, to 1.4m long, becoming white toward the petiole. Petiole is glaucous, to 60cm long, the rachis, glaucous below, yellow above to 4.4m long, extending beyond the terminal leaflets by about 15cm. Leaflets, about 70 per side, are pendulous, acuminate, more or less with parallel margins to 1m long and 4cm wide, reducing in length toward the apex, green above, lighter green below with a light green mid rib above, scaled on the veins and with ramenta on the mid rib below.

The infrafoliar inflorescence is glabrous, spreading to about 90cm long, branched to 3 orders, the 1st order branches at an angle greater than 45° to the rachis, ultimate rachillae to 40cm long, flexuous becoming pendulous in fruit. Prophyll and peduncular bract greenish. The glabrous peduncle is short, the rachis much longer. Triads are spirally arranged on the basal ½, above a bare basal portion, paired and solitary male flowers distally. Male flowers have valvate petals, 6 stamens and a short trifid pistillode. Female flowers are smaller than the males. Fruits are globose, to 6mm in diameter, with subapical stigmatic remains, black at maturity with persistent perianth. Endocarp is thin and fragile, the seed to 4mm in diameter with a basal embryo.

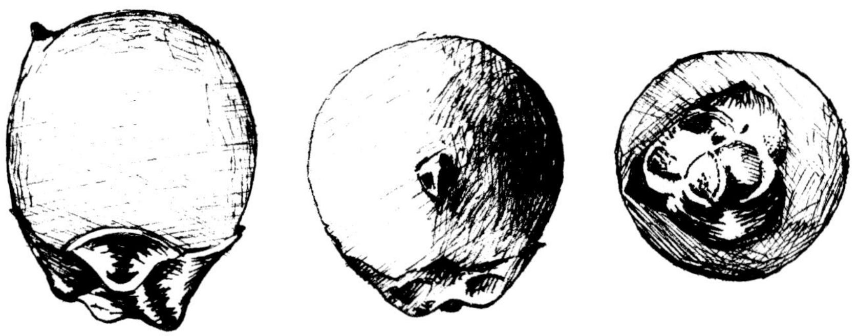

Clinostigma harlandii. Fruit (three views)(six times natural size)

Figure 49: Upper. Inflorescence in bract and released inflorescence of *Lepidorrachis mooreana,* Lord Howe Island (CT).
Lower left. *Lepidorrachis mooreana* on Lord Howe Island (CT).
Lower right. Seedling of *Lepidorrachis mooreana* (CT).

Clinostigma exorrhizum (H. Wendl.) Becc. 1935

syn. *Areca exorrhiza*
Clinostigma seemannii
Kentia exorrhiza
Exorrhiza thurstonii
Exorrhiza wendlandiana
Exorrhiza smithii
Clinostigma thurstonii
Clinostigma smithii
Clinostigmopsis seemannii

Locale, Habitat and Ecology:

Fiji, on the islands of Viti Levu, Ngau, Vanua Levu and Taveuni, as an emergent palm in very moist forests on ridge crests and steep slopes at from 230-1,200m. This species is most common in its upper limits, areas of very heavy rainfall, over 6,000mm per annum on Taveuni, and with persistent mist and cloud cover. The famed tangimauthia, *Medinilla waterhousei,* a semi-epiphytic climber with translucent red flower bracts, is found in the vicinity at the higher altitudes on Taveuni and Vanua Levu.

Description:

A tall palm with a closely ringed, whitish to brown trunk to 20m, 35cm in diameter with brown, spiny stilt roots to 3m long at the base. Mosses and epiphytic ferns thickly cover the trunk's surface. Arching to pendulous leaves number to about 12, are to 5m long. The crownshaft, to 1.2m long, is light green/ glaucous, more so toward the petiole. The broad petiole, to 60cm long, extends into an arching rachis and beyond the apical leaflets by about 15cm. Leaflets are dark green, acuminate, pendulous, numbering to about 100 per side, are prominently ribbed with thickened margins, reducing in length toward the apex, to 1m long and 4cm wide.

The infrafoliar inflorescence is horizontal, becoming pendulous, to 65cm long, branched to 3 orders with more or less parallel rachillae, white at anthesis, becoming green in fruit. The prophyll encloses the peduncular bract, both beaked. Triads are borne almost throughout the thin rachilla, the male flowers cream/orange with 6 stamens and a short trifid pistillode. The females smaller with recurved stigmas. The fruit is ovoid, to 6mm long by 4mm in diameter, translucent red at maturity, with apical stigmatic remains. The endocarp is thin, fragile, operculate, the seed with an elongate hilum and basal embryo.

Clinostigma exorrhizum. Fruit and endocarp (five times natural size)

Figure 50: Upper left. *Clinostigma harlandii* at high altitude on Ambrym, Vanuatu (PC).
Upper right. *Clinostigma exorrhizum* at about 1100m asl on Des Voeus Peak, Taveuni, Fiji.
Lower. Herbarium specimen of *Clinostigma harlandii* in Tagabe Herbarium (PVV), Port Vila, Vanuatu.

Figure 51: Following page. Upper left. Trunks of *Carpoxylon macrospermum;* note the expanded bases and the abundant lichen.
Upper right. Leaf of *Carpoxylon macrospermum.*
lower. Fruit and seed of *Carpoxylon macrospermum.*

Figure 52: Opposite following page. *Cyphokentia macrostachya* at low altitude in the Riviere Bleue basin, southern New Caledonia (AD).

CARPOXYLON H.A. Wendl. & Drude 1875

A monotypic genus endemic to Vanuatu. Its position within Iguanurinae is relatively isolated, but appears most closely related to *Clinostigma.* The large fruit with a ridged boney endocarp is very distinctive.

Carpoxylon macrospermum H.A. Wendl. & Drude 1875

Locale, Habitat and Ecology:
Vanuatu, originally described from Aneityum (Anatom), where it is now most probably extinct due to the cultivation practices of the indigenous peoples, but now known from only a few individuals in supposed cultivation. One substantiated report, upon which the complete generic and specific descriptions are based (Dowe & Uhl 1989), is on Espiritu Santo, whilst unsubstantiated but reliable reports suggest that it is cultivated on the islands of Ambai and Erromango. The original location was noted as at higher elevations, but the known palms occur at sea-level in occasionally inundated riverine habitat in a more or less exposed situation.

Fruit bats (flying fox) have been observed consuming the epicarp and thin fleshy mesocarp of fruits on the tree, whilst coconut crabs ingest the softish endosperm of the fallen fruits after crushing them with their powerful claws.

This species was thought lost to botany until rediscovered in 1987 (Dowe 1988) during the preparatory field study for this work. It appears very adaptable to cultivation.

Description: A tall palm to 18m with a trunk to 25cm in diameter, enlarged at the base to 50cm in diameter. Leafscars are whitish, internodes elongate but closer spaced distally. Leaves number about 12, are ascending and strongly arched. The crownshaft is elongate, columnar, slightly narrowing toward the petioles, light glossy green, glabrous. The petiole is stout, to 25cm long, narrowing into the rachis to 3.75m long, extending beyond the apical leaflets into a flexible tip of about 12cm long, all similarly covered as the crownshaft. Leaflets, about 70 per side, are linear acuminate with an irregular bifid apex, glabrous green above, lighter green beneath with punctate scales, ascending erectly at mid leaf, becoming horizontal toward the base and the apex, mid rib prominent, more so above, margins thickened, with less prominent lateral ribs, lacking ramenta below. Leaflets are strongly reduplicate at attachment, measuring from 1.5cm to 3.2cm wide and from 36cm to 1.22m long.

The infrafoliar inflorescence is held erectly, branched to 3 orders, glabrous, green in fruit, ultimate rachillae to about 40cm long. The prophyll, to about 70cm long and 8cm wide is two keeled, glabrous, dark green, the peduncular bract tubular, exerted and tapered to a woody point, inserted about 5cm above the attachment of the prophyll, a second peduncular bract, similar to the former, inserted 5cm above, is enclosed within. The peduncle is stout, elliptical in cross- section, about 14cm long with scars of incomplete bracts below the rachis, which is also stout, to about 36cm long. Low bracts subtend the primary branches, which are angled. Rachillae are stout, rigid, about 5mm thick, tapered toward the apex. Triads, green in bud, are distantly and spirally arranged on the lower one- quarter of the rachilla, about 1cm apart, paired or solitary males distally, one male flower is distal and one lateral to the female. Male flowers are about 4.5mm high, irregular, the sepals almost as long as the petals, 6 stamens, with anthers about 2mm long, the pistillode about as long as the stamens. Female flowers, about 6mm high, are irregular with 3 fleshy stigmas. The fruit is obovoid to elliptical, 6cm long by 3.5cm in diameter, tapered toward the base, prominent stigmatic remains subapical, red at maturity. The epicarp is smooth, the mesocarp thickly fibrous, the endocarp boney, not easily separated from the mesocarp, with thick longitudinal ridges, the operculum large, circular to spatulate. The seed is ellipsoidal, to 4cm long and 2.5cm in diameter, with abundant raphe fibres, extending laterally, the endosperm homogenous with a small central cavity, the embryo basal.

Figure 53: *Carpoxylon macrospermum,* south Espiritu Santo, Vanuatu.

CYPHOKENTIA Brongn. 1873

A monotypic genus endemic to New Caledonia, most closely related to *Moratia* also from New Caledonia, as well as possessing fruit similar to that of *Brongniartikentia.*

Cyphokentia macrostachya Brongn. 1873
syn. *Cyphokentia robusta*
Clinostigma macrostachyum
Clinostigma robustum
Dolichokentia robusta

Locale, Habitat and Ecology:
New Caledonia, commonly over southern, central and north-eastern La Grande- Terre, as a semi-emergent species in moist forests on micaschists and serpentine soils on acid and ultrabasic rocks at from 150-900m. It is most common on the slopes of Mounts Kanala and Arago and near Aoupine between 400-900m and is the most prevalent species of the Riviere Bleue Basin at 200-300m.

Description:
A medium palm with a greyish white trunk, closely ringed, to 10m and 15cm in diameter, enlarged at the base with adventitious roots. The spreading, arching leaves number about 10. The crownshaft is to 1m long, light green with a whitish wax. The petiole, to 17cm and the arcuate rachis to 2.5m are glabrous. The stiff leaflets number 25-30 per side, are glossy dark green above, paler below with pale brown ramenta beneath. The leaflets are from 1cm to 4.6cm wide by 14cm to 1.35m long, the basal leaflets being reduced to lorae.

The inflorescence is much branched with thin pendulous rachillae to 80cm long. The prophyll is silver glaucous, waxy white, to 60cm long by 15cm wide, enclosing the peduncular bract. Triads are borne throughout the rachillae. Male flowers are rose coloured in bud, with valvate petals, 12 stamens and an angled, apically flat pistillode equal to the stamens in length. The females are much smaller, to about 6mm wide in fruit. The fruits are red at maturity, ellipsoidal to 2.2cm long and 1.5cm in diameter, with sublateral stigmatic remains, an ovoid operculate endocarp, thickish with a basal keel. The seed, to 1cm long and 6mm wide lacks the basal keel, but has a basal hilum and embryo.

Cyphokentia macrostachya. Fruit and endocarp (two views)(twice natural size)

MORATIA H.E. Moore 1980

A monotypic genus endemic to New Caledonia, very closely related to *Cyphokentia* also of New Caledonia. Basic differences include stamen numbers, 6 in this genus, globose fruit (ellipsoidal in *Cyphokentia*) and a ridged endocarp (smooth in *Cyphokentia*).

Moratia cerifera H.E. Moore 1980
syn. *Basselinia glabrata*

Locale, Habitat and Ecology:
New Caledonia, in the north-east of La Grande-Terre, in moist forests on acid rocks, most common on the eastern slopes of Mt. Panie at 400-1,600m but also found as a single specimen at 10m growing with *Alloschmidia glabrata, Veillonia alba, Chambeyronia macrocarpa, Basselinia gracilis, Cyphosperma balansae* and other species which are also more common at the higher altitudes. The occurrence

Figure 54: Villagers holding an infructescence of *Carpoxylon macrospermum,* southern Espiritu Santo, Vanuatu.

of palms at near sea-level which are usually found at moderate to high altitudes may be partly explained by a parallel situation which occurs with some of the palm taxa in north Queensland. Species, such as *Archontophoenix* sp. 'Mt.Lewis', *Laccospadix australasica* and *Oraniopsis appendiculata,* all of which are commonly found at high altitudes, sporadically occur at sea or near sea-level in sites where the cooler mountain air descends protected gullies, thus creating microclimates which resemble those to be found at the higher altitudes (pers.com. A.K.Irvine). A similar situation may occur with *Alsmithia longipes* on Taveuni in eastern Fiji.

Description:
A tall palm with a brown trunk to 20m high and 18cm in diameter, prominently ringed, lacking adventitious roots. The spreading, arching leaves number about 9. The crownshaft is orangish, with a thick, waxy, white covering, to 94cm long. The petiole is green, brown lepidote beneath, brown floccose above, to 22cm long extending into a rachis to 1.7m long, similarly covered. The stiff leaflets number about 30 per side, directed forward, dark green above and below, the mid rib yellow/green. Ramenta are on the mid rib and lateral veins beneath. Leaflets measure from 17cm to 96cm long and 5mm to 5cm wide.

The infrafoliar inflorescence is orangish/green to reddish/grey, glaucous, broader than long with ultimate rachillae to 62cm long, flexuous, arching to semi- pendulous. The short peduncle, to 5cm long, is elliptical in cross-section, 6cm wide by 1.4cm thick, the rachis is to 43cm long with a branching peduncular base, thence twice branched. The orangish prophyll, to 60cm long, encloses the peduncular bract. Triads are borne almost throughout the rachilla, paired and single males distally, subtended by rounded, spreading bracts. The male flowers are red in bud, with valvate petals twice as long as the sepals, 6 stamens and a pistillode equal in length to the stamens. The female flowers are slightly smaller, red in bud. The fruits, to 1.3cm high and 1.4cm wide are ellipsoidal/subglobose with lateral stigmatic remains. The mesocarp is fleshy, to 1mm thick. The endocarp has a spatulate operculum, is lightly patterned, thickish and vitreous. The seed is globose, broader than high, to 6mm high by 8mm wide. The raphe branches from the base, the embryo basal.

Moratia cerifera. Fruit (two views) and endocarp (two and one half times natural size)

BASSELINIA Vieill. 1873

A genus of eleven species endemic to New Caledonia. This is a diverse genus, divided into two not markedly distinct groups, but Section TALOUA of seven species is generally composed of tall non-caespitose species with fruit exceeding 1cm in diameter, crownshafts generally green and leaflets always once ribbed, whilst Section BASSELINIA with four species is generally composed of small caespitose species with fruit less than 1cm in diameter, crownshafts in colours other than green and leaflets united withmore than one rib or undivided but pinnately ribbed. This genus is closely related to genera such as *Clinosperma, Moratia,* and *Cyphokentia* of New Caledonia and the widespread *Clinostigma. Basselinia gracilis* is the type species.

Figure 55: *Moratia cerifera* in habitat at 600m asl on Mt. Panie', north-eastern New Caledonia (O)

Key to the Species

Stems single, medium to tall palms
Leaf regularly pinnate
Fruit usually exceeding 1 cm in diameter
Crownshafts predominantly green

Section TALOUA

Leaf with 15-25 leaflets per side
Rachillae essentially glabrous

B. humboldtiana

Rachillae essentially velutinous
Bracts subtending triads obscured by trichomes

B. porphyrea

Bracts subtending triads not obscured by trichomes
Leaf with 30-45 leaflets per side

B. iterata

Petals of lighter colour than sepals

B. velutina

Petals of darker colour than sepals

B. sordida

Petals of similar colour

Rachillae densely tomentose
Bracts subtending triads obscured by trichomes
Triads widely spaced
Fruit black

B. tomentosa

Triads closely spaced
Fruit red
Inflorescence massive

B. favieri

Stems usually caespitose, small to medium palms
Leaf undivided, bijugate, irregularly divided or regularly pinnate
Fruit usually less than 1 cm in diameter
Crownshaft predominantly not greenish but lilac, orange, red, brown, purple to purple-black
Generally confined to serpentine soils

Section BASSELINIA

Usually caespitose, smallish palms
Fruit globose
Leaf lepidote beneath
Leaf undivided and pinnately ribbed or bijugate
Rachillae glabrous
Ramenta on leaflet below sparse or absent

B. vestita

Leaf not lepidote beneath
Rachillae scaly
Triads with prominent separating bracts, male flowers with sepals more than ½ as long as petals

B. deplanchei

Leaf irregularly or regularly divided (rarely bijugate)
Rachillae tomentose
Triads lacking prominent separating bracts, male flowers with sepals less than ½ as long as petals more or less obscured by hairs

B. gracilis

Usually single stem, tallish palms
Fruit bilobed
Ramenta prominent on leaflet below
Male flowers with sepals not obscured by hairs

B. pancheri

Basselinia humboldtiana (Brongn.) H.E. Moore 1984
syn. *Cyphokentia humboldtiana*
Clinostigma humboldtiana

Locale, Habitat and Ecology:
New Caledonia, where it is locally abundant in 2 localities in the southern sector of La Grande-Terre, on Mt.Humboldt and Montagne des Sources in moist forest on serpentine soils at 800-1,000m. It usually grows in association with *Basselinia porphyrea* in both localities at 900-1,000m and with *Actinokentia huerlimannii* at 850-1,000m on Mt.Humboldt.

Description: A medium palm to 15m with a trunk to 10cm in diameter, prominently ringed. Leaves, about 8, are ascending in a compact crown. The crownshaft is dark green, brown scaled, slightly bulging, the petiole to 6cm long and the rachis smooth, to 1.6m long. Leaflets, about 20 per side, are leathery, prominently ribbed, glossy green above, paler beneath, held erect, from 8mm to 6cm wide and from 57cm to 1.1m long.

The infrafoliar inflorescence, to about 90cm long, is essentially glabrous, green, red toward the base, with ultimate rachillae to 50cm long. Subtending bracts are low and rounded. The prophyll does not fully encircle the peduncle at the base. The peduncle is stout, about 6cm long, 5cm wide and 2.5cm thick. The triads are arranged in horizontal lines in sunken pits, subtended by a small bract. Flowers are reddish in bud. The male flowers, to 2mm high, have sepals and petals almost of equal length, 6 stamens, the pistillode as long as the stamens, a 3 angled apex. Female flowers larger to 4mm high. The fruit is obovoid, dark brown at maturity, 1cm long by 8mm in diameter, stigmatic remains almost apical. The epicarp is smooth, mesocarp thin, endocarp thin and fragile, operculum rounded and basal, the seed globose.

Basselinia porphyrea H.E. Moore 1984

Locale, Habitat and Ecology:
New Caledonia, where it is confined to Mt.Humboldt and Montagne des Sources at 900-1,000m in moist forest on serpentine soils. It grows in association with *Basselinia humboldtiana.*

Description:
A small palm to 5m with a closely ringed trunk to 10cm in diameter. The leaves, erect to slightly arching, number about 7. The crownshaft, to 52cm long, is slightly bulging, greenish purple with grey scales. The petiole, to 20cm, and the rachis, to 1.2m are green. Leaflets, about 20 per side, are erect and leathery, directed forward, dark green above and below, from 5mm to 4cm wide and from 28cm to 83cm long.

The inflorescence, about 65cm long, is stiff, densely covered in velutinous tomentum, light to ferric red. Both the prophyll, which does not fully encircle the peduncle at attachment, to 34cm long and 8cm wide and the peduncular bract to 38cm long, are glabrous. The peduncle and rachis are short, the rachillae ultimately to 23cm long. Triads are borne to almost the full length of the rachilla in horizontal lines, subtended by bracts which are hidden in the dense tomentum. The male flowers, to 3mm high, have sepals almost the same length as petals, stamens 6, the pistillode rounded, as long as the stamens. Female flowers about 5mm high. The fruits are subglobose, to 1.1cm in diameter, the stigmatic remains subapical. The mesocarp is about 1mm thick, the seed globose.

Basselinia porphyrea. Fruit and endocarp (twice natural size)

Basselinia iterata H.E. Moore 1984

Locale, Habitat and Ecology:
New Caledonia, where it is restricted to the eastern ridges of Mt. Ignambi in the north-eastern sector of La Grande-Terre, in moist forests on schistose soils at about 1,000m in stunted lower montane forest.

Description:
A medium palm to 7m with a trunk to 8cm in diameter with prominent flat leaf scars. Leaves, about 6, are ascending. The crownshaft, about 60cm long, is green with dark scales. The petiole, to 44cm long, and the rachis, to 1.2m long, are similarly coloured and scaled. Leaflets, about 16 per side, are lax, directed forward, glossy green above, paler beneath, with ramenta on the base of the mid rib below. They measure from 7mm to 8cm wide and from 26cm to 1.2m long.

The infrafoliar inflorescence, to 45cm long, is spreading, densely covered in short soft scales. The peduncle is short to 5cm, the rachis elongate to 17cm, bracts subtending branches low. Triads are subtended by low bracts, not obscured by scales. The male flowers, to 3mm high, have sepals almost as long as the petals, 6 stamens with yellow anthers, and orange pistillode. The female flowers are larger, to 4.5mm high, reddish brown at receptivity. The fruits are subglobose, about 1cm in diameter, black at maturity, with subapical stigmatic remains.

Basselinia iterata. Fruit and endocarp (twice natural size)

Basselinia velutina Becc. 1920

Locale, Habitat and Ecology:
New Caledonia, where it is distributed in disjunct locations, Mt. Dogny in the south-western sector and Mt. Panie in the north-east, in moist forest on acid rocks at 400-1,600m. Mt. Panie has one population at 500m and another at 1,200m to the summit (1,628m) whilst the population on Mt. Dogny is at 400-1,000m.

Description:
A medium palm to 12m with a smooth trunk to 15cm in diameter. Leaves, about 11, are arching. The crownshaft, to 1m long, is brown, grey red, red or slate grey, with dense floccose scales. The petiole, to 35cm, and the rachis, to 3m, are similarly scaled. Leaflets, about 43 per side, are crowded toward the base, green above, lighter green below, the mid rib prominent with brownish ramenta on the mid rib below. They measure from 4mm to 6cm wide and from 45cm to 1.35m long, the basal leaflets reduced to lorae.

The infrafoliar inflorescence, to over 1m long, is velutinous/tomentose, shortly peduncled, with

---

Figure 56: Upper. *Basselinia tomentosa* at 1000m asl on Mt. Nakada, central east New Caledonia (O)
Lower left. Infructescence of *Basselinia favieri*, New Caledonia; note the crowded fruit and thick covering of tomentum (RF).
Lower right. *Basselinia pancheri* at low altitude on the slopes above Lake Yate near Port Perignon, New Caledonia (AD).

Figure 57: Following page. *Basselinia humboldtiana* in habitat (LHBH).

Figure 58: Opposite following page. *Basselinia porphyrea* at 900m asl on Montagne des Sources, south New Caledonia (O)

branches to 75cm long and ultimate rachillae to 55cm long, branches and rachillae more or less pendulous. The prophyll, either completely or incompletely encircling the peduncle at attachment, is brown scaled, the peduncular bract exceeding the prophyll, with an undeveloped bract sometimes present. Triads subtended by low bracts. Male flowers, to 5mm high, have sepals of a dark brown about equal to the lighter brown petals, stamens 6, pistillode orange, trifid at the apex. Female flowers about the same size as the males. Fruits are subglobose, about 1.1cm long by 8mm wide, black at maturity, stigmatic remains lateral or lower.

Basselinia sordida H.E. Moore 1984
syn. *Pseudobasselinia alba*

Locale, Habitat and Ecology:
New Caledonia, being restricted to a limited area in the western sector of La Grande-Terre around Mt.Boulinda and Me Maoya at 1,000-1,400m in moist forest on serpentine rocks. Other palms which occur in this area include *Cyphosperma balansae* and *Clinosperma bracteale* but at a lower altitude.

Description: A medium palm to 9m with a trunk to 15cm in diameter and with prominent leaf scars. The leaves number about 10. The crownshaft, to 70cm long, is shallowly split opposite the petiole, dark green with brown/grey scales. The broad petiole, to about 18cm long, and the arching rachis to almost 3m long are covered in floccose scales beneath. The leaflets, about 32 per side, are dark green, basally crowded, measuring from 6mm to 5cm wide and from 24cm to 87cm long, the basal leaflets reduced to lorae.

The infrafoliar inflorescence, to about 60cm long and 90cm wide, is erectly spreading, divaricately branched, whitish with scales, ultimate rachillae to 40cm long. Triads are subtended by low bracts. The male flowers are green on the inside, tawny-green outside, about 4mm high and 3mm wide, the sepals about the length of the petals, 6 stamens with purple anthers, the pistillode nearly as long as the petals. Female flowers brown at anthesis, slightly larger than the males. Fruits are subglobose, about 1cm in diameter with subapical stigmatic remains.

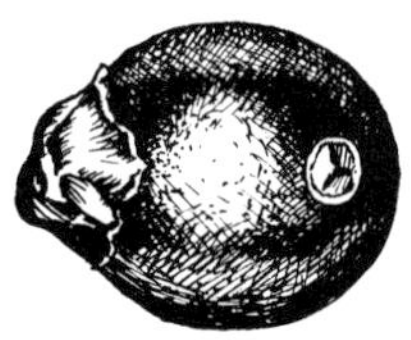

Basselinia sordida. Fruit (twice natural size)

Basselinia tomentosa Becc. 1920

Locale, Habitat and Ecology:
New Caledonia, confined to the ridges near the summit of Mt.Nakada in the southern sector of La Grande-Terre, as an emergent palm in moist forest, on schistose soils at an altitude of 1,000-1,100m. Other palms which occur within the vicinity include *Clinosperma bracteale* at a similar altitude and *Actinokentia divaricata* at lower altitudes. This species is one of the more robust of the genus and distinguished by the thickly tomentose inflorescence and grey scaled crownshaft. It has been confused with *B.favieri* of Mt.Panie, but differs in basic anatomy as well as the leaves being shorter, the inflorescence and flowers being smaller and the trunk being much taller and robust.

Description:
A tall palm with a prominently ringed trunk to 25cm in diameter. Leaves number about 9, are erect, the green crownshaft to about 1m long is covered with grey scales. The petiole, to about 5cm long and the rachis, are covered with brown lepidote scales. Leaflets number about 32 per side, are rigid, dark green with prominent ramenta on the mid rib below. They measure from 8mm to 5cm wide and from 53cm to 1m long.

The infrafoliar inflorescence is large and spreading to about 80cm long and wide, densely covered with whitish tomentum, once branched with ultimate rachillae to 42cm long. The prophyll, which does not fully encircle the peduncle at attachment, is greenish/brown to 19cm wide, the peduncular bract brown scaled. Triads are subtended by bracts which are buried within the tomentum, widely spaced. The flowers are purplish, the males to 2mm high, the sepals almost as long as the petals, 6 stamens, an angled pistillode with a winged apex is as long as the stamens. The female flowers are about 3.5mm high. The fruit is subglobose, black at maturity, about 1cm long with subapical stigmatic remains.

Basselinia favieri H.E. Moore 1984
syn. *Basselinia tomentosa*

Locale, Habitat and Ecology:
New Caledonia, where it is locally abundant but confined to Mt. Panie at an altitude of 300-500m. This area, in the north-east sector of La Grande-Terre, receives the highest annual rainfall in New Caledonia, in excess of 4,000mm per annum. The forests are moist and cool. Other palms which grow in the vicinity include *Lavoixia macrocarpa, Chambeyronia lepidota, Veillonia alba, Basselinia gracilis* and *B. velutina.* This palm is outstanding by virtue of the massive, thickly tomentose inflorescence which is by far the largest in the genus.

Description:
A medium palm to 10m with a trunk to 15cm in diameter, expanded at the base and with obvious but not raised rings. Leaves number about 9, are ascending and spreading. The crownshaft is large, to 1.2m long, green, covered with grey scales, the petiole, about 35cm long, and the rachis, to about 3.5m long, have reddish suffusions. Leaflets, about 45 per side, are green, acute, with ramenta on the mid rib below. They measure from 1.2cm wide to 1.55m long, the basal leaflets reduced to lorae.

The infrafoliar inflorescence is thickly tomentose, large, to about 2m across, divaricately branched with ultimate rachillae to 48cm long and 1.1cm thick. Triads are very closely placed, with bracts and braceole obscured by tomentum. Male flowers are about 4mm high, sepals about ½ the length of the petals, stamens 6, the pistillode longer than the stamens. Female flowers are about 5mm high with strongly recurved stigmas. Fruit is subglobose, about 1cm long, red at maturity, with subapical stigmatic remains. The epicarp is smooth, the mesocarp thin, the endocarp thin and fragile.

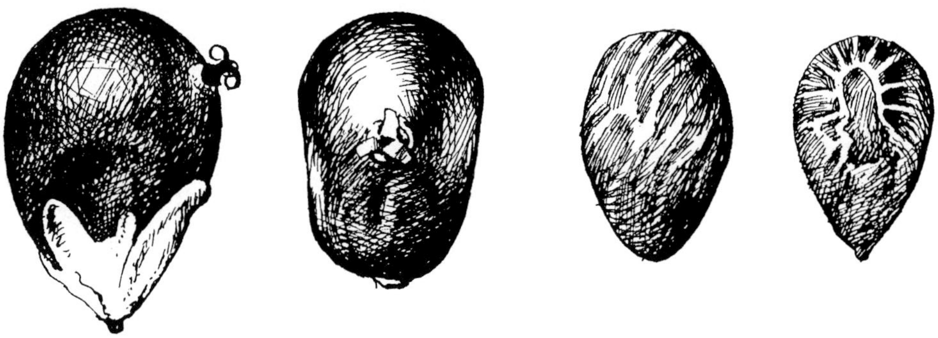

Basselinia, favieri. Fruit (two views) and endocarp (two views)(four times natural size)

Basselinia vestita H.E. Moore 1984

Locale, Habitat and Ecology:
New Caledonia, being confined to Me Ori in the central-west sector, north-west of Dogny, as an understorey palm in shaded, moist forest on serpentine soils at 900-1,000m. *Clinosperma bracteale* often grows nearby.

Description:
A small palm, usually caespitose but occasionally as a single trunk to 1.5m. The leaves, about 6, are

Figure 59: Following page. *Basselinia velutina* in habitat, New Caledonia (LHBH).

Figure 60: Opposite following page. *Basselinia favieri* at 300m asl on Mt. Panie', north-east New Caledonia (O)

usually undivided, bijugate or sparingly but irregularly divided. The crownshaft is to 16cm long, reddish lepidote, the petiole short and the rachis, to about 20cm long, are similarly scaled. The leaves are dark green above, densely grey vestite below, with ramenta on the mid ribs below. When divided, they measure 3cm to 4.5cm wide and from 34cm to 44cm long, and undivided about 6cm wide and 42cm long.

The infrafoliar inflorescence is generally lepidote but with glabrous rachillae ultimately to 25cm long. The prophyll, which does not fully encircle the peduncle at attachment, is blackish lepidote, to 13cm long, the peduncular bract similarly lepidote to 17.5cm long. Triads have variable, spreading to acute bracts. The flowers are red, the males to 1.5mm high, sepals almost as long as the petals. Female flowers are about 1.8mm high. Fruits are globose, black at maturity, about 7mm in diameter with apical stigmatic remains.

Basselinia vestita. Fruit (natural size)

Basselinia deplanchei (Brongn. & Gris) Vieill. 1873
syn. *Kentia deplanchei*
Cyphokentia surculosa
Cyphokentia deplanchei
Clinostigma deplanchei
Clinostigma surculosa
Microkentia surculosa
Microkentia deplanchei
Microkentia schlechteri
Basselinia surculosa

Locale, Habitat and Ecology:
New Caledonia, where it is of uncommon occurrence but of widespread distribution in the southern one-third of La Grande-Terre, on serpentine soils in association with ultrabasic rocks at altitudes of 500-1,000m, in both open forests or natural clearings as a semi-understorey component or somewhat stunted heliophyte.

This is one of the few clump forming species in the entire South-West Pacific. Its taxonomy has been confused with high altitude forms of *B.gracilis*, but readily distinguished by unobscured male sepals, prominent bracts subtending the triads, closely spaced triads and choice of an exposed habitat.

Description:
A small caespitose palm to 4m and trunks with a diameter of 5cm, usually covered with persistent leaf-bases, but where they have fallen, the exposed trunk having prominent nodes. Leaves number from 5 to 10, are ascending and spreading, usually undivided, but if divided, with one-ribbed leaflets. The crownshaft, slightly bulging, is about 18cm long, red to black, covered with deciduous scales. The petiole, to 15cm, and the rachis, to 50cm, are similarly scaled. The leaflets are leathery, united and pinnately ribbed or regularly pinnate and once ribbed, and occasionally bijugate, dark green above, lighter green beneath, lepidote on the ribs with sparse ramenta on the base of the mid rib below. Leaflets measure from 5mm to 1.5cm wide to 15cm to 52cm long.

The infrafoliar inflorescence is dark green/red, thrice branched with ultimate rachillae to 40cm long. The prophyll, which does not fully encircle the peduncle at attachment, is lepidote, the peduncular bract similar. Rachillae are rigid, with lepidote scales, the triads raised above the scales. The male flowers are dark purple, to 1.6mm high, sepals about the length of the petals, 6 stamens, the pistillode white, not as long as the stamens. Fruits are subglobose, 8mm long by 6mm wide, black at maturity, with subapical stigmatic remains. The seed is globose, about 4mm in diameter.

Basselinia deplanchei. Fruit and endocarp (natural size)

Basselinia gracilis (Brongn. & Gris) Vieill. 1873
syn. *Kentia gracilis*
Cyphokentia billardieri
Cyphokentia eriostachys
Cyphokentia gracilis
Clinostigma eriostachys
Clinostigma gracilis
Kentia eriostachys

Microkentia billardieri
Microkentia gracilis
Basselinia eriostachys
Basselinia billardieri
Basselinia heterophylla
Microkentia heterophylla

Locale, Habitat and Ecology:
New Caledonia, throughout most of La Grande-Terre, in moist forests on a variety of substrata, but more frequently on schistose or serpentine soils, from near sea level to the summit of the highest mountain (Mt.Panie 1,628m). It reaches its highest population density at altitudes about 900m, areas with high annual rainfall. At lower altitudes, it is a wispy trunked, finely leafed understorey component, whilst at higher altitudes it chooses more exposed positions being almost a true heliophyte, somewhat stunted, with short infloresences and undivided leathery leaves.

In evolutionary terms, this genus represents a unique development generically and specifically. Regarding the caespitose habit, *Basselinia* is unique within the South-West Pacific and is also one of the few caespitose genera within Iguanurinae (e.g. others include *Iguanura, Heterospathe* and *Sommieria).* In a way this species somehow typifies the concept of the New Caledonia palms, being smallish, an understorey component, having a dramatically coloured crownshaft, tomentose inflorescence and small fruits with prominent stigmatic remains.

Description:
A small to medium variably palm, usually caespitose, very infrequently single trunked, to 8m with a trunk diameter of 7mm to 10cm. Nodes are prominent and the internodes scaled. Leaves, from 5 to 12, ascending but descending with age. The crownshaft is short to elongate, to 45cm, coloured in pink, rose, red, crimson, reddish black, slate black to grey. Suffusions of these colours extend into the thin petiole, to 30cm long, and into the rachis to 1.2m long. Leaflets, if the leaf is divided, number about 15 per side, or if united with pinnate mid ribs or bijugate. They are glossy, green above, lighter green below, ramenta lacking or very sparse on the mid rib(s) below. They measure from 4mm to 14cm wide and from 7.5cm long to 74cm long.

The infrafoliar inflorescence is lightly tomentose, dark red to brown, spreading, with ultimate rachillae to 32cm long. Triads are closely spaced, in depressions which lack subtending bracts, the flowers dark red. Male flowers are about 1.6mm high, sepals about ½ the length of the petals, usually obscured by tomentum, 6 stamens with a white pistillode not as long as the stamens. Female flowers are about 2.5mm high with strongly recurved stigmas. The fruit is globose, about 6mm in diameter, black at maturity, subapical stigmatic remains with a persistent perianth. The seed is globose to about 4mm in diameter.

Basselinia gracilis. Endocarp (three views)(three times natural size)

Figure 61: Following page. *Basselinia vestita* in habitat (LHBH).

Figure 62: Opposite following page. *Basselinia sordida* in habitat (LHBH).
Lower left. *Basselinia deplanchei* with mature infructescence (O)
Lower right. *Basselinia pancheri* with inflorescence, at low altitude in the Riviere Bleue basin near the 'refuge hut', southern New Caledonia (AD).

Basselinia pancheri (Brongn. & Gris) Vieill. 1873

syn. *Kentia pancheri*
Cyphokentia pancheri
Clinostigma pancheri
Microkentia pancheri
Nephrocarpus schlecteri

Locale, Habitat and Ecology:

New Caledonia, widespread throughout La Grande-Terre, but more commonly in the southern sector, where it occupies moist to dryish forest on serpentine soils at 200-1,200m (most prevalent at 800-1,000m), occasionally in shaded situations, but generally as a semi-emergent in forest with a light broken canopy. Specimens from higher altitudes display a stunted habit with smaller inflorescence and narrow, evenly pinnate leaves, but the flowers and fruits remaining remarkable consistent with others from lower altitudes. The kidney shaped and laterally attached fruits are unique within the genus.

Description:

A small to medium palm to 10m with a trunk to 10cm in diameter, bright green/grey with prominent nodes and elongate internodes. Leaves, about 7, are erectly ascending, unevenly pinnate. The crownshaft is slightly bulging, to about 55cm long, coloured in red, orange, purple-black, slate grey to brown, usually yellowish into the petiole, with deciduous lepides. The petiole, to 25cm long, and the rachis to 1.3m long, are similarly scaled. Leaflets are irregularly united, pinnately ribbed, irregularly spaced on the rachis, dark green above, lighter beneath, with usually abundant ramenta on the ribs below. Leaflets measure from 4mm to 16.5cm wide and from 16cm to 1.35m long.

The infrafoliar inflorescence is held erect to horizontal, to 60cm long, 4 times branched with lepidote rachillae ultimately to 30cm long. The prophyll is pale floccose, to 26cm long by 5cm wide, not fully encircling the peduncle, the peduncular bract exerted to 32cm long, similarly covered. Triads are well spaced, subtended by rounded bracts. Flowers are reddish, the males to 1.5mm high, sepals ½ or more the length of the petals, 6 stamens, white, the pistillode white, about as long as the stamens. Female flowers, to 5mm high, have prominent stigmas. Fruits are kidney shaped to bilobed, laterally attached, the stigmatic remains lateral, black at maturity. The seed is kidney shaped and bilobed, brown.

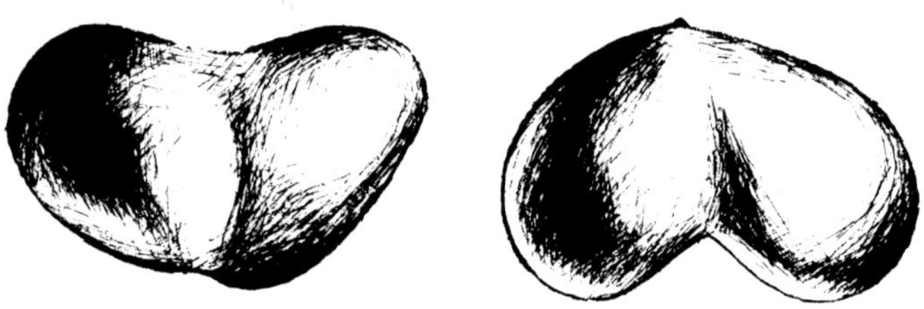

Basselinia pancheri. Endocarp (two views)(three times natural size)

CYPHOPHOENIX H.Wendl. ex J.D. Hooker 1883

A genus of two species endemic to New Caledonia, disjunctly distributed, occurring on soil types of different structures. Appears most closely related to *Campecarpus,* also of New Caledonia. *Cyphophoenix elegans* is the type species.

Key to the Species

Medium palm with arching leaves with about 26 leaflets per side
Ramenta glossy
Sepals of male flowers brown
Fruit ovoid, seed with sparsely branched raphe fibres
Occurs on acid rocks
C. elegans

Large palm with horizontal leaves with about 40 leaflets per side
Ramenta dull
Sepals of male flowers green
Fruit ellipsoidal, seed with much branched raphe fibres
Occurs on calcareous rocks
C. nucele

Cyphophoenix elegans (Brongn. and Gris) H.Wendl. ex Salomon 1887
syn. *Kentia elegans*

Locale, Habitat and Ecology:
New Caledonia, confined to the extreme north-eastern sector of La Grande-Terre as a semi-emergent species in moist forests bordering streams on soils over acid rocks at an elevation of 20-550m. It is most common near the creeks of Amos Gap and the Mayavetch Basin, the latter locality being shared with *Mackeea magnifica.*

Description:
A medium palm with a prominently ringed trunk to 10m and 20cm in diameter. The arching leaves number about 8. The crownshaft is olive green/grey, to 90cm long, extending into a short petiole to 17cm and an arching rachis to 2m, both of which are brown lepidote beneath. The stiffly erect leaflets number to 27 per side, are dark green above, paler beneath with shining ramenta, measure from 2mm to 3.5cm wide by 20cm to 1.1m long. The basal leaflets are reduced to lorae.

The short peduncled inflorescence is broader than long, 1.3m wide by 70cm long. The prophyll, covered in light brown scales, is 33cm long and 6cm wide, the peduncular bract, similarly scaled, exceeds the prophyll and is 36cm long. The inflorescence is well branched with ultimate rachillae to 65cm long. Triads are borne on the basal ½, subtended by liplike bracts, the males to 4mm high, with brown sepals, valvate petals, 6 stamens and a columnar pistillode about the same length as the stamens. Female flowers are to 6mm high with prominent bracteoles. The fruits are ovoid, 2cm long by 1.4cm in diameter, with apical stigmatic remains. The endocarp is laterally grooved, operculate, the seed, to 1.7cm long by 7mm wide, has an apical hilum and sparsely branched raphe fibres.

Cyphophoenix elegans. Fruit (two views)(twice natural size)

Cyphophoenix nucele H.E. Moore 1976
syn. *Microkentia sp.*

Locale, Habitat and Ecology:
New Caledonia, where it is restricted to a small area on Lifou Island, on raised limestone in coastal rainforest at 60m. This is the only New Caledonian species which does not occur on La Grande-Terre.

Description:
A tallish palm with a prominently ringed, basally swollen trunk to 12m and 20cm in diameter. The horizontally held leaves number about 8. The light green crownshaft is pale lepidote, to 60cm long, extending into a short petiole to 7cm long and a rachis to 2m long. The glossy, dark green leaflets number about 40 per side, are paler beneath with prominent dull ramenta on the mid rib below. They measure from 2.5mm to 3.6cm wide and 46cm to 78cm long, the basal leaflets developing into lorae.

The inflorescences are broader than long, short peduncled, glabrous in fruit with ultimate rachillae to 26cm long. Triads are borne on the basal ½ of the rachilla, the male flowers are greenish to 4.5mm high with 6 stamens and a yellow pistillode, the females are to 6mm high with prominent bracteoles. The

Figure 63: Following page. Upper left. Brilliantly coloured crownshaft of *Basselinia gracilis,* Riviere Bleue basin, New Caledonia (AD).
Upper right. *Basselinia gracilis,* in habitat, Riviere Bleue basin, New Caledonia (AD).
Lower. Crowns of *Campecarpus fulcitus,* New Caledonia (RF).

Figure 64: Opposite following page. *Cyphophoenix elegans* in habitat, northern New Caledonia (RF).

fruit is red at maturity, ellipsoidal, to 2cm long by 1.2cm in diameter, the seed similarly shaped to 1.1cm long by 7mm in diameter, with much branched raphe fibres.

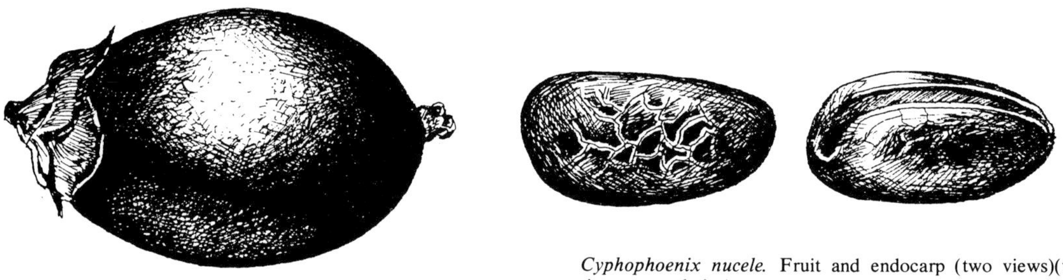

Cyphophoenix nucele. Fruit and endocarp (two views)(three times natural size)

CAMPECARPUS H.Wendl. ex Becc. 1920

A monotypic genus endemic to New Caledonia, most closely resembling *Cyphophoenix* also of New Caledonia.

Campecarpus fulcitus (Brongn.) H.Wendl. ex Becc. 1920
syn. *Kentia fulcita*
Cyphophoenix fulcita

Locale, Habitat and Ecology:

New Caledonia in the southern sector of La Grande-Terre at altitudes of 100- 800m, most commonly occurring as scattered individuals on Mt.Koghi and in the catchment areas of Riviere Bleue and Riviere Blanche, as a semi-emergent species in moist forests on serpentine soils.

Description:

A medium palm to 7m with a prominently ringed green/brown trunk to 26cm in diameter, with brown prop roots to 50cm long profusely produced at the base of the trunk. The spreading to erect leaves number from 6-10. The elongate crownshaft, to 1m long, is light green to grey, densely covered with woolly tomentum. The short petiole, to 15cm, and the rachis, to 2.55m long, are covered with brown scales. Leaflets number about 35 per side, dark shiny green above, paler beneath, leathery, with recurved margins and ramenta present on the mid rib below. They measure from 2cm to 7.5cm wide and 21cm to 1.32m long, the basal leaflets reduced to lorae.

The infrafoliar inflorescence is glabrous green, held horizontally, with stout pendulous rachillae ultimately to 45cm long. The glabrous prophyll does not fully encircle the peduncle at the base, is split below, about 40cm long by 6cm wide, and does not enclose the peduncular bract which is 60cm long and 10cm wide. Triads are borne in the lower two-thirds of the rachilla, subtended by lip-like bracts. Male flowers are rosy green in bud, 6-7mm high with valvate petals, sepals half their length, stamens 6 and a pointed pistillode shorter than the stamens. Female flowers are about 1cm high. The fruits, narrowly ovoid with a curved apex and apical stigmatic remains are about 2.75cm long by 1.3cm in diameter. They are black at maturity with a smooth epicarp, fleshy mesocarp, fragile endocarp with a

Figure 65: *Campecarpus fulcitus* in habitat, New Caledonia (RF).

Figure 66: Following page. Upper left. *Campecarpus fulcitus* at low altitude in the Riviere Bleue basin near the 'refuge hut', southern New Caledonia (AD).
Upper right. Stilt roots of *Campecarpus fulcitus* (AD)
Lower. Inflorescence (in bud) of *Campecarpus fulcitus* (AD)

Figure 67: Opposite following page. Crown of *Veillonia alba* with inflorescences in bud and young fruit; note the imcompletely encircling prophyll at the base of the inflorescence (RF).

longitudinal ridge and basal operculum. The seed is ellipsoidal, 1.5cm long by 8mm in diameter, with an elongate hilum and basal embryo.

Campecarpus fulcitus. fruit (twice natural size)

VEILLONIA H.E.Moore 1978

A monotypic genus endemic to New Caledonia, most closely related to *Burretiokentia* from which it differs by having the fruit surface minutely papillate, the pistillode longer than the stamens and having a waxy cover on the lower crownshaft, upper trunk and inflorescence bracts.

Veillonia alba H.E.Moore 1978
syn. *Pseudocyphosperma alba*

Locale, Habitat and Ecology:
New Caledonia, confined to the north-east sector of La Grande-Terre on the Mt.Panie massif and nearby areas, in moist forest on gneissic and schistose soils generally at 200-600m but with an isolated occurrence at near sea-level in an area where other higher altitude species are also to be found (see *Moratia*).

Description:
A medium palm to 7m with a grey/brown trunk to 12cm in diameter with long internodes and the upper trunk whitish with a waxy covering. Leaves, about 10, are spreading. The crownshaft, about 1.2m long, shallowly split opposite the petiole, is whitish grey with dense dark red floccose scales, scattered at the base, becoming very dense toward and into the petiole, covered with a whitish wax. The petiole, to 60cm long and the arching rachis to 2.8m long are brown lepidote below. The leaflets, about 45 per side, are held horizontal, acuminate, rigid with a prominent mid rib, green above with lepidote scales below on the mid rib and lateral ribs and brown ramenta to 8mm on the mid rib below. Leaflets measure from 5mm to 6.2cm wide and from 24cm to 1.2m long.

The infrafoliar inflorescence is stiffly spreading to 60cm long, minutely papillate, becoming green in fruit. The prophyll, to 25cm long, does not fully encircle the peduncle at the base, is split below, the peduncular bract to 40cm long and 15cm wide, both bracts with a waxy surface. The peduncle, to 10cm long, is elliptical in cross-section. The rachis and rachillae, ultimately to 40cm long, are flexuous and glabrous with bracts to 2.5cm subtending the lower branches. Triads are spirally arranged on the lower ½ of the rachilla, solitary or paired males on the narrowing distal portion, bracts subtending flowers prominent and acute. Male flowers are creamy white, about 3.5mm long, sepals much shorter than the valvate petals, stamens 6, pistillode capitate and longer than the stamens. Female flowers are to 6mm high with sepals almost as long as the petals. The fruit is ovoid, about 1.6cm long and 1.3cm in diameter, brown at maturity with stigmatic remains subapical. The epicarp is papillate, the mesocarp soft and lightly fibrous, the endocarp is fragile, sculptured, longitudinally ridged and irregularly protuberant, the operculum basal subtending a dorsal groove. The seed is similarly shaped and ridged with an elongate hilum and basal embryo.

Veillonia alba. Fruit (two views) and endocarp (natural size)

Figure 68: *Burretiokentia vieillardii* in habitat, New Caledonia (RF).

BURRETIOKENTIA Pichi-Sermolli 1955

A genus of two species endemic to New Caledonia. It is closely related to *Veillonia* (also of New Caledonia) from which it differs in having a smooth rather than papillate epicarp, a pistillode shorter than the stamens and lacking a waxy coat on the upper trunk and lower crownshaft. Otherwise, the close similarity of these two genera may warrant their combination. *Burretiokentia vieillardii* is the type species.

Key to the Species

Inflorescence glabrous and about as broad as is long
 Fruit globose
 Endocarp with mid ridges less prominent than lateral ridges. — *B. vieillardii*

Inflorescence tomentose and about twice as broad as is long
 Fruit ellipsoidal
 Endocarp with mid ridges more prominent than lateral ridges — *B. hapala*

Burretiokentia vieillardii (Brongn. & Gris) Pichi-Serm. 1955
syn. *Kentia vieillardii*
Cyphosperma vieillardii
Rhynchocarpa vieillardii
Cyphokentia robusta

Locale, Habitat and Ecology:
New Caledonia, where it is a common widespread species on La Grande-Terre as an emergent sometimes conspicuous component in moist forests on serpentine soils and schists from 200-1,000m. It is the most common palm around the Amieu Gap and is very common in the mountains of the east coast becoming scarcer in the Mt.Koghi region in the south.

Description:
A tall palm to 18m with a green, prominently ringed trunk to 20cm in diameter. Prickly adventitious roots are often present. The horizontal twisted leaves number about 10, with a compact to elongate slightly bulging green crownshaft to about 1m, with reddish/brown scales. The short petiole, to 20cm, and the rachis, to 2.4m, are similarly scaled. The acute leaflets number about 30 per side, densely scaled beneath and measuring from 3mm to 7cm wide by 14cm to 93cm long, basally reduced to lorae.

The infrafoliar inflorescence is compact, with a stout peduncle, creamy-green to dark green, glabrous, three-times branched, with ultimate rachillae to about 40cm long. The short prophyll, to about 30cm, is rusty red and the peduncular bract, exceeding the prophyll to 34cm long, is covered with brown tomentum. Triads are borne on the lower two-thirds of the rachilla, subtended by low rounded bracts, the flowers whitish. The male flowers, to 5mm high, have short pedicels, valvate petals, 6 stamens with white anthers and a pistillode not as high as the stamens. Female flowers, to 4mm high have bracteoles to 2.5mm high. The globose or occasionally pear-shaped fruits measure to 1.5cm long by 1cm wide, with a rough epicarp, reddish/brown at maturity and almost apical stigmatic remains. The endocarp is symmetrically sculptured, the mid ridges less prominent than the lateral ridges, and with a basal operculum. The seed is similarly shaped, to 1.2cm long by 9mm wide and 5mm thick, with a basal embryo.

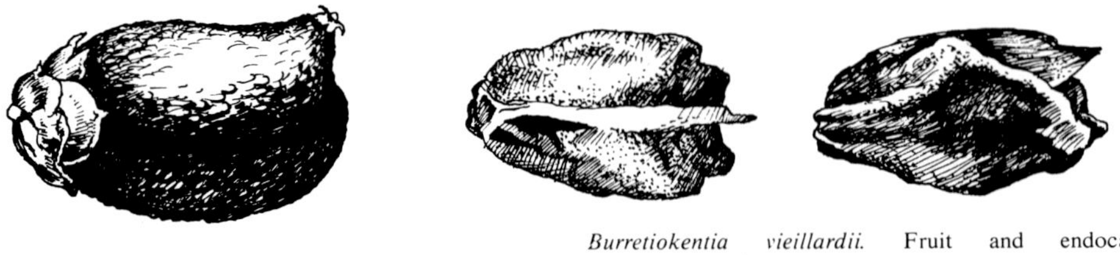

Burretiokentia vieillardii. Fruit and endocarp (two views)(natural size)

Figure 69: Mature fruit of *Burretiokentia vieillardii,* New Caledonia (RF).

Burretiokentia hapala H.E.Moore 1969

Locale, Habitat and Ecology:
New Caledonia, where it is confined to the extreme northern sector in both moist and relatively dry forests on limestone and schistose soils at 50-400m. This emergent palm is locally abundant in the moist Parari Gap area and in the relatively hot and dry area between Koumac and Ouegoa.

Description:
A medium palm to 10m with a prominently ringed, green trunk to a diameter of 20cm. The spreading leaves number from 9-10, with a compact, occasionally elongate crownshaft with a slight bulge, olive green with brownish scales. The short petiole, to 15cm, and the ascending rachis, to 2.4m long, are similarly scaled. The leaflets, about 40 per side, are held horizontal, with an acute to oblique apex, glossy green above, lighter green beneath and strongly reduplicate at attachment. The mid rib has membraneous ramenta below. Leaflets measure from 6mm to 5cm wide and from 27cm to 90cm long.

The infrafoliar inflorescence is wider than broad, with erect to pendulous branches and thickly tomentose rachillae to 35cm long. The prophyll is about 22cm long and the peduncular bract about 31cm long. The prophyll incompletely encircles the peduncle at attachment. Male flowers are green, to 5.5mm high, petals slightly longer than the sepals which are ciliolately margined, 6 stamens and a pistillode shorter than the stamens. Female flowers are about 6mm high. The fruits are ovoid to ellipsoidal, becoming reddish/brown at maturity, to 1.6cm long and 9mm in diameter with stigmatic remains almost apical. The endocarp is sculptured, the mid ridges more prominent than the lateral ridges, and a rounded operculum at the base. The seed is similarly shaped, 8mm long by 6.5mm in diameter.

PHYSOKENTIA Becc. 1934

A genus of seven species widespread from New Britain, Solomon Islands, Vanuatu to Fiji. Closely related to *Cyphosperma, Burretiokentia* and *Veillonia* with which it shares sculptured endocarp and seed and an incompletely encircling prophyll split below, and also sharing with *Alsmithia* and *Lavoixia* the sculptured endocarp but not their completely encircling prophylls. The former lacks a crownshaft. *Physokentia tete* is the type species.

Key to the Species

Fruit exceeding 2.5cm long
- Inflorescence broader than long
  - Endocarp minutely ridged and indented
  - Seed with 4 lateral ridges
  - Operculum rounded
  - Bracteoles and pedicels glabrous
    - Vanua Lava, Vanuatu — *P. tete*

Fruit less than 2.5cm long
- Inflorescence longer than broad
  - Endocarp not minutely ridged or indented
  - Seed with 2 lateral ridges
    - Rachilla deep rose
    - Flowers with red petals and red/black sepals
      - Viti Levu, Ngau (?) Fiji — *P. rosea*
  - Operculum angled
  - Bracteoles and pedicels barbate
    - Rachilla ivory
    - Flowers with white or pink flushed petals
      - Taveuni, Fiji — *P. thurstonii*

Figure 70: Upper left. Crownshaft and inflorescences of *Physokentia rosea,* at about 1000m asl on the eastern ridge of Mt. Lomalangi, northern Viti Levu, Fiji.
Upper right. Part of inflorescence of *Physokentia rosea* with male flowers at anthesis.
Lower. Leaf of *Physokentia rosea.*

Physokentia tete (Becc.) Becc. 1934
syn. *Cyphosperma tete*

Locale, Habitat and Ecology:
Vanuatu, on the islands of Vanua Lava, Pentecote and Erromango uncommonly as an understorey palm in moist forest on volcanic rocks at 500-1,000m. On Erromango, it grows in *Agathis* forest on ferrallitic soils (Schmid 1970). *Clinostigma harlandii* is usually found growing with this species, although more commonly and widespread. It is uncertain if this species has stilt roots as do all the other species in the genus. Collections and descriptions are only fragmentary.

Description:
A medium palm with a trunk diameter of about 7cm. Leaves regularly pinnate. Leaflets obliquely truncate, about 60cm long and 6cm wide, with a prominent mid rib and thickened secondary ribs, dull green above, paler beneath with brown scales on the ribs below.

The infrafoliar inflorescence is broader than long, branched to three orders, initially scaled but becoming glabrous in fruit. The peduncle is short, the rachis angled, the ultimate rachillae flexuous to 15cm long. The peduncular bract is brown and thin. Bracts subtending the triads are prominent and acute, with narrow bracteoles surrounding the female flowers. Male flowers are angled, to 6mm long by 3mm wide, with ciliolately margined sepals about one-third the length of the petals, the pistillode is columnar, enlarged at the apex. Female flowers are ovoid/conic. The fruit is globose, about 2.5cm long by 2.3cm in diameter, with subapical stigmatic remains. The epicarp is smooth, the mesocarp fleshy with few fibres. The endocarp is thick with prominent keels and two prominent ridges on each side. The surface is minutely ridged and indented, the operculum rounded. The seed is similar to the endocarp in shape, 1.7cm long by 1.5cm in diameter.

Physokentia rosea H.E.Moore 1966
syn. *Goniocladus petiolatus (?)*

Locale, Habitat and Ecology:
Fiji, endemic to Viti Levu, uncommonly in the mountains toward the centre of the island above 750m, as an emergent species forming loose colonies in mossy cloud forest on slopes and ridges.

Description:
A medium palm to 8m with a dark green, usually moss covered trunk to 10cm in diameter, prominent, irregular leafscars and stilt roots about 1m long and 4cm thick, reddish with a papery tip emerging from the trunk base. Leaves, about 9, are spreading, slightly curving, and stiff. The crownshaft, about 50cm long and thickened at the base, is olive green with deciduous floccose scales, pink flushed toward and into the petiole. The greenish petiole, to 22cm long, and the reddish rachis, to 2m long, are both densely lepidote, more so below. Leaflets, about 35 per side, are once ribbed, acute with decurved margins, dull green, the mid rib lepidote above and the mid rib and lateral veins irregularly scaled below. Leaflets measure from 6mm to 4cm wide and from 16cm to 67cm long.

The infrafoliar inflorescence is held erect to horizontal, about 50cm long, glabrous, red in bud becoming green in fruit. The peduncle, to 9cm long, is narrow, branched, elliptical in cross-section at the base, the lower branches angled, the angled rachillae becoming terete, flexuous, ultimately to 24cm long. Prophyll and peduncular bract similar, thin, pinkish to white basally. Triads are spirally arranged to about ½ the length of the rachilla, paired or solitary male flowers distal, subtended by acute bracts. Male flowers, about 5mm high, deep rose in bud, sepals about one-third the length of the petals, widely spreading at anthesis, stamens 6 with white anthers, pistillode trifid apically shorter than the stamens. Female flowers are about 5mm high, red/black, strongly imbricate sepals and red petals. The fruit, globose, to 2cm in diameter, is shiny black at maturity. The epicarp is smooth, with subapical stigmatic remains, mesocarp is fleshy, the endocarp angled, keeled and longitudinally ridged, the operculum roundish, the seed similarly shaped, to 1.2cm long by 1cm wide and thick, with a basal embryo.

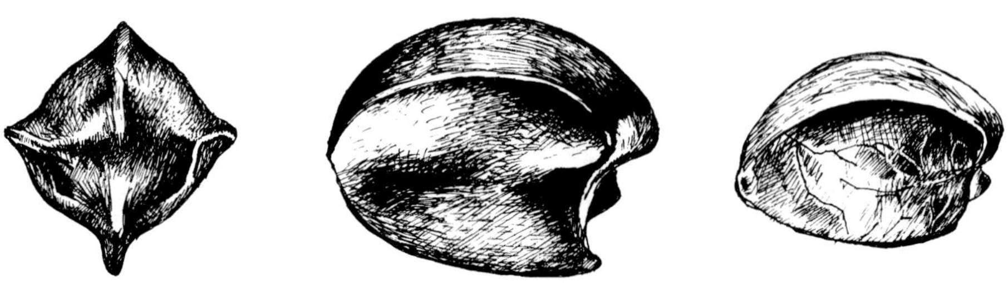

Physokentia rosea. Endocarp (two views) and seed (natural size)

Physokentia thurstonii (Becc.) Becc. 1934
syn. *Cyphosperma thurstonii*
Goniosperma vitiense
Goniosperma thurstonii

Locale, Habitat and Ecology:
Fiji, on Vanua Levu and Taveuni, uncommonly in moist rainforest on slopes at 600-1,000m. On Taveuni, this species grows in association with *Clinostigma exorrhizum* and *Cyphosperma trichospadix* in an area which receives upward of 7,000mm of rain per annum.

Description:
A medium palm to 7m with a green trunk to 10cm in diameter, prominent leaf scars and prickly, brown, occasionally branched, stilt roots at the base. Leaves, about 9, are spreading. The crownshaft, to 55cm long, is olive/dark green with brown floccose scales. The petiole is rounded, to 37cm long, and the rachis, to 1.7m, are densely scaled. Leaflets, about 27 per side, dark green glabrous above, lighter green below, have decurved margins, brown scales at the base of the mid rib above and small brown scales on the mid rib, minor ribs and marginal veins below. Leaflets measure 3mm to 8cm wide and from 20cm to 75cm long.

The infrafoliar inflorescence is held erect to horizontal, the peduncle branched, branches and rachillae angled, the rachillae stiffish, ivory coloured, ultimately to 20cm long. Prophyll, about 17cm long, and the peduncular bract, to 19cm long, are thin and whitish. Triads are borne to about ½ the rachilla length, paired male flowers distal, subtended by reflexed, acute bracts. Male flowers are pinkish/white, about 6mm high, sepals about one-quarter the length of petals, stamens 6, pistillode angled, not as high as the stamens. Female flowers about 4.5mm high with narrow, membraneous bracteoles. The fruit are globose, about 2.3cm long by 2.1cm in diameter, black at maturity, with apical stigmatic remains. The mesocarp is thick, whitish, the endocarp thick with 4 sharp angles, 2 laterally and an upper and lower keel, the operculum angled. The seed is similarly shaped to 1.7cm long by 1.3cm thick, raphe branches radiating from the upper keel, the embryo basal.

CYPHOSPERMA H.Wendl. ex J.D.Hooker 1883

A genus of four species, one in New Caledonia, one in Vanuatu (un-named) and two in Fiji, closely related to but distinct from *Burretiokentia, Physokentia* and *Veillonia* and more closely to *Alsmithia,* a genus which may prove to be con-generic with *Cyphosperma. Cyphosperma balansae* is the type species.

Cyphosperma, Burretiokentia, Physokentia and *Veillonia,* and *Alsmithia* and *Lavoixia* are a group of genera united by sculptured endocarps or seeds, the first four genera by an incompletely encircling prophyll split below, *Alsmithia* to *Cyphosperma* by their elongate peduncle and lack of crownshaft, and *Lavoixia* to *Physokentia* and *Veillonia* by the sculptured endocarp and well developed crownshaft. With further study, some of these genera may be combined, e.g. *Veillonia* with *Burretiokentia, Lavoixia* with *Physokentia* and *Alsmithia* with *Cyphosperma.*

Figure 71: Following page, Upper. Crown of *Cyphosperma balansae* with inflorescences at 10m asl at Pouai Creek, north-east New Caledonia (O).
Lower left. Stilt roots of *Physokentia rosea.*
Lower right. Crown of *Cyphosperma tanga;* note the elongate peduncle of the inflorescence and the undivided leaf blade, above Vuninatambua Creek, northern Viti Levu, Fiji.

Figure 72: Opposite following page. Fruit of *Lavoixia macrocarpa,* in habitat at 500m asl on Mt. Panie', north-eastern New Caledonia (O).
Lower left. Newly emerged leaf of *Cyphosperma tanga* held by Eliki Namaialau, Forestry Assistant from Navai, Viti Levu, Fiji.
Lower right. Glossy brown trunk of *Cyphosperma tanga.*

Key to the Species

Crownshaft absent
Inflorescence interfoliar
Peduncle greater than 1m long
Leaf undivided except irregularly toward the apex
Mid rib and veins scaled on the upper surface
Fruit oblong/ellipsoidal to 1.3cm long
Taveuni, Fiji
C. tanga

Leaf divided regularly throughout
Mid rib and veins lacking scales on the upper surface
Fruit globose to 1.2cm in diameter
New Caledonia
C. balansae

Peduncle less than 70cm long
Espiritu Santo, Vanuatu
C. sp.

Crownshaft present
Inflorescence infrafoliar
Fruit oblong-ellipsoidal to 2cm long
Vanua Levu, Taveuni, Fiji
C. trichospadix

Cyphosperma tanga H.E.Moore 1977
syn. *Taveunia tanga*

Locale, Habitat and Ecology:
Fiji, endemic to Viti Levu, in the vicinity of Navai, Dromodromo Creek in the north and Serua Province, inland from Namboutini, in the south, where is forms scattered groups in moist dense forests at altitudes from 600-900m, as an understorey species on slopes with little soil and large volcanic rocks.

Description:
A medium palm with a chocolate brown, undulately ringed trunk to 5m and 15cm in diameter. The erectly held leaves number to about 12. Leafbases are clasping, light green, extending into a short petiole to 15cm and a rachis to 2m. Petiole and rachis are light green and glabrous. The leaf is usually undivided except for irregular divisions in the apical 20cm, dark green above, paler beneath, with scales on the ribs below.

The reddish inflorescence, to 2m long, is initiated as interfoliar, becoming pendulous with a flexuous peduncle to 1.2m long. The prophyll incompletely encircles the peduncle at attachment above the base, the peduncular bract is inserted well above the prophyll. Both bracts are persistent, papery and slow to decay. The rachis is to 80cm long and the rachillae of an equal length. Triads are borne throughout the rachilla, the flowers being greenish, to 2.2mm high. Male flowers have valvate petals, 6 stamens and a thick prominent pistillode. Female flowers are larger than the males, roundish with 3 recurved stigmas. The fruit is oblong-ellipsoidal, 1.3cm long, yellowish at maturity. The endocarp is sculptured with an adaxial ridge and a basal operculum. The seed is similarly shaped.

Cyphosperma balansae (Brongn.) H.Wendl. ex Salomon 1887
syn. *Cyphokentia balansae*

Locale, Habitat and Ecology:
New Caledonia, where it is endemic to La Grande-Terre, being found uncommonly in the mid-central and western sectors as a semi-emergent species in moist forests on schists and peridotites at altitudes of 10-1,200m. It is most common on Mts. Arago and Aoupinie between 400-1,000m whilst small colonies are to be found in the Boghen Valley and on the Boulinda Massif at 1,200m.

Description:

A medium palm to 15m with a dark green, scarred trunk to 15cm in diameter. The leaves, to about 30, are more or less ranked. The crownshaft is indistinct, with clasping leafbases to 1m long with a brown base becoming pinkish toward the petiole. The petiole is stout, to 40cm, green with brown tomentum, the rachis similarly covered, extending to 3m. Leaflets are light green above, lepidote beneath, numbering to about 35 per side with pale brown ramenta on the mid rib below, they measure from 2cm to 12cm wide and from 28cm to 1.1m long.

The reddish inflorescence is initiated as interfoliar, becoming pendulous with a flexuous peduncle to 2m long. The prophyll incompletely encircles the peduncle at attachment, splitting below, attached about 3cm above the base, whilst the peduncular bract is inserted about 6cm above the attachment of the prophyll, both bracts are persistent, papery and slow to decay. The peduncle branches at about 2m, with ultimate rachilla to 22cm long. Bracts subtending the branches are low, rounded to acute to 5mm, whilst those subtending the triads are similarly shaped but only 1.5mm high. Triads are borne throughout the rachillae, the male flowers rosy/brown in bud, to 4mm high, with valvate petals, 6 stamens, pale anthers and a prominent, orange pistillode. The female flowers are larger than the males, to 4.5mm high, rounded with 3 recurved stigmas. The fruit is globose, to 1.2cm in diameter, red at maturity and with apical stigmatic remains. The endocarp is sculptured, with lateral ridges and a large operculum at the base. The seed is similarly shaped, to 8mm wide and 9mm long with a linear hilum and basal embryo.

Cyphosperma balansae. Fruit (natural size)

Cyphosperma trichospadix (Burret) H.E.Moore 1977
syn. *Taveunia trichospadix*

Locale, Habitat and Ecology:

Fiji, on the islands of Vanua Levu and Taveuni, occurring on the slopes of Mt.Mariko on the former and on the latter near the central mountain summits, particularly above Somosomo and near Lake Tangamauthia, as an understorey species on slopes and ridges of very moist forests on volcanic rocks at altitudes of 600-1,200m.

Description:

A medium palm to 7m with a bright green/brown trunk to 8.5cm in diameter. Regularly divided leaves measure to 1.8m long. The crownshaft is prominent. The petioles are long.Leafletsnumber about 25 per side, are acute to acuminate with a prominent mid rib and less prominent lateral ribs.

The infrafoliar inflorescence, to 1.5m long, is initiated as interfoliar, at first erect,becoming pendulous in fruit. The peduncle is to 65cm long, the rachis of similar length. Ultimate rachillae are to 60cm long. The prophyll is persistent, fibrous in age, more so toward the apex. The peduncular bract is inserted toward the apex of the prophyll. Triads are borne throughout the rachilla, the flowers greenish to 3mm high. Male flowers have valvate petals, 6 stamens and a prominent pistillode. The female flowers are slightly larger. The fruits are oblong-ellipsoidal to 2cm long with apical stigmatic remains. The epicarp is smooth, the endocarp sculptured and ridged with a seed of similar shape.

Cyphosperma sp. Vanuatu

Locale, Habitat and Ecology:

Vanuatu, on Espiritu Santo, in the north of Cumberland Peninsula above Pialorai River on Voutmele Ridge, as an understorey species in dense moist forest on volcanic soils at an altitude of 900-1,100m. This species is known from a small colony of about 7 individuals. Only reported to botany in 1979, this species was at first thought to represent *C. balansae,* a species indigenous to New Caledonia (N.Uhl) but this determination has been contested by the collector. Further argument for it being a new species

has been put forward by J.-M. Veillon upon examination of material in the Herbarium at Noumea.

Description:
From Collection Notes; MORAT (Philippe) 6488. Espiritu Santo, 3rd August 1979. North Santo - on slopes of Voutmele, 900-1,100m. A slender tree 5-6m, growing in a small colony of 6/7 individuals. Annular scars very marked and close, stripe (trunk) very cylindrical. Leaves 1.5m, divided into leaflets. Inflorescence pendulous, 3/4 from within the axis of the leaves, more than 1m in length, fertile branches (rachillae) 4/5, 40-50cm in length at the extremity.

ALSMITHIA H.E.Moore 1982

A monotypic genus endemic to Taveuni in Fiji. The relationship to *Cyphosperma* is close, but differences include a prophyll which completely encircles the peduncle at attachment, rachillae with a bare basal portion, elongate fibres on the endocarp: all characters which may delimit species rather than genera.

Alsmithia longipes H.E.Moore 1982

Locale, Habitat and Ecology:
Fiji, on Taveuni, in very moist forests on steep slopes, ridges and rocky stream edges, the main population at 300-500m, with an isolated group known at sea- level near Lavena on the south-east coast. Taveuni is of recent volcanic formation, being one of the wettest islands in the entire region, with rainfall approaching 10,000mm per annum in the vicinity where this species occurs.

Description:
A smallish palm to 4m, with an irregular, closely noded brown trunk to 7.5cm in diameter. The leaves, about 12, are ascending, lacking a crownshaft, with short clasping leafbases, with brown scales, splitting opposite the petiole. The petiole is long, to 80cm, scaled, becoming rounded in cross section, the rachis similarly scaled, arching erect to 2.8m long. Leaflets, about 35 per side, are dark green above, lighter below, linear acute, rigid, a prominent mid rib and thickened lateral ribs, lacking scales and ramenta below, from 5mm to 6cm wide and from 28cm to 75cm long.

The interfoliar inflorescence is large, to 1.5m long, green to salmon pink with deciduous brown tomentum. The prophyll, to 55cm long and 5cm wide, is green with brown lepidote scales, persistent, the peduncular bract inserted about 22cm above the prophyll, similarly scaled and persistent. Both bracts become salmon pink with age. The peduncle, to 60cm long, is thick, angled, as are the lower branches. The rachillae, ultimately to 60cm long, are rounded, basally bare, flexuous, becoming apically thin and more or less pendulous in fruit. Triads are borne spirally to about ½ the length of the rachilla, bracts low and inconspicuous. The male flowers are creamy/white to 4.5mm long, the petals valvate, about 3 times as long as sepals, stamens 6 with yellow anthers, the pistillode orange, slightly shorter than the stamens. Female flowers about 5.5mm high. The fruit is ellipsoidal, about 3.5cm long by 2.3cm in diameter, bright crimson at maturity,

with apical stigmatic remains. The epicarp is smooth, the mesocarp about 4mm thick with oblique fibres, the endocarp sculptured, thin and fragile, ridged and sculptured with fibres extending from the lateral ridges and keels, operculum rounded and basal, the seed angled with a prominent hilum, smooth, raphe branches spreading from the apex, the embryo basal.

LAVOIXIA H.E.Moore 1978

A monotypic genus endemic to New Caledonia. Its relationships are obscure, but possibly lie with other genera with sculptured endocarps and seeds, particularly *Physokentia* and *Veillonia* and less so with *Burretiokentia*. The large fruit with lateral stigmatic remains and the absence of anther connectives (didymous) are remarkable.

Figure 73: Upper left. *Alsmithia longipes* in habitat, north-eastern Taveuni, Fiji.
Upper left. Leaf of. *Alsmithia longipes*.
Lower. *Alsmithia longipes* inflorescence in bud; note the persistant prophyll and peduncular bract.

Lavoixia macrocarpa H.E.Moore 1978

Locale, Habit and Ecology:
New Caledonia, where it is known from only five individuals confined to the eastern side of Mt. Panie' in moist forest at 500m. It is a semi-emergent species growing amongst rocks on a steep slope. Its extinction appears imminent due to apparent lack of germination of mature fruits both in the wild and in cultivation.

Description:
A medium palm to 12m with a brown trunk to 20cm in diameter with an expanded base. The spreading arching leaves number about 7. The crownshaft is bulging, dark green with thick grey scales tapered toward the petioles. The olive green petiole, to about 26cm long, rounded in cross-section, extends into a green rachis to 2.3m long. Leaflets, about 30 per side, are rigid, ascending, dark green above, with a yellowish mid rib, dull green beneath with brown ramenta on the mid rib below. Leaflets are about 2.3cm to 5.7cm wide and 35cm to 70cm long.

The infrafoliar inflorescence is spreading, about 80cm long with upward curving branches and rachillae, becoming semi-pendulous in fruit. The prophyll, 35cm long and 9cm wide, and the peduncular bract, exceeding the prophyll, to 42cm long, are both lepidote. The inflorescence is much branched with ultimate rachillae to 27cm long, branches and rachillae subtended by low bracts. Triads are borne only on the basal portion, sunken, solitary and paired males distal, subtended by prominent rounded bracts. Male flowers are red in bud, about 2mm high, sepals much smaller than the valvate petals, stamens 6, anthers lacking a connective, pistillode with a flat cap exceeding the stamens in length. The females are reddish, about 5mm wide. The fruit, almost globose, about 4cm long by 3.4cm in diameter, has lateral stigmatic remains. The epicarp is smooth, the mesocarp thick, the endocarp thin, laterally compressed with a sharp median ridge, sculptured and pitted in appearance, the operculum basal. The seed is similarly shaped, to 2cm long and 5mm thick, the embryo basal.

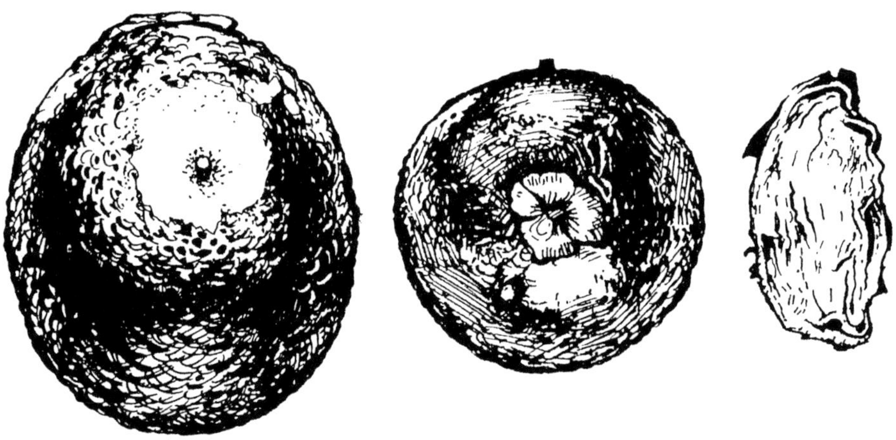

Lavoixia macrocarpa. Fruit (two views) and endocarp (one and one-half times natural size)

GONIOCLADUS Burret 1940

An obscure genus known only from the original description of a palm collected in 1937 by St.John (1838). Although *Physokentia,* a genus found at the type locality appears similar, *Goniocladus* cannot be assuredly placed under synonymity until it has been recollected and fully studied. The most recent search, though unsuccessful, was undertaken by H.E.Moore Jr. in 1964.

Goniocladus petiolatus Burret 1940

Locale, Habitat and Ecology:
Fiji, on the island of Viti Levu, originally collected in the vicinity of Rairaimatuku Plateau in Nandronga/Navosa-Naitasiri Province in rainforest at 1,158m.

Description:
A small palm to 7m with a trunk diameter of 20cm. A prominent crownshaft extends into a channeled petiole to 15cm and a rachis to 1m long, densely covered in powdery indument. The leaflets, 24 per side are acute to acuminate, once ribbed with ramenta on the mid rib below.

The infrafoliar inflorescence is thrice branched to 30cm long. The prophyll, papery, to 14cm long, is exceeded by the peduncular bract to 16cm long. Rachillae, to 20cm long, are pointed. Triads are spirally arranged throughout, the male flowers with 6 stamens and a capitate pistillode. The fruit (immature) is ovate with apical stigmatic remains.

BUTIINAE Saakov

Subfamily: Arecoideae
Tribe: Cocoeae

Butia	*Allagoptera*
Lytocaryum	*Cocos**
Jubaea	*Polyandrococos*
Parajubaea	*Syagrus*
Jubaeopsis	* Indigenous genus

Description:
Small to very large, understorey to emergent; solitary or clump forming; non-sheathing, pinnately leaved; unarmed or armed; monoecious, pleonanthic; peduncular bract woody; endocarp with three pores.

Distribution:
From equatorial to temperate; sea-level to low altitudes; rainforest, moist forests, seasonally dry forests, savannah, coastal forests, sand dunes. South-East Africa, equatorial to temperate South America, pan-tropical island foreshores, West Indies and Central America.

Notes:
Most of the genera in this SUBTRIBE occur in South America, only *Cocos* being represented in the South-West Pacific. It is a ubiquitous genus, occuring north of the Tropic of Capricorn. Origin of this genus is unresolved and open to speculation. (See Origins)

Because of the widespread distribution of this genus within the South-West Pacific, a map is not included.

COCOS L. 1783

A monotypic genus, considered native to the western Pacific, but most likely occurring in the South-West Pacific as a consequence of the migration of man to the region. It is most closely related to *Syagrus,* a genus occurring in South America.

Cocos nucifera L. 1783

Locale, Habitat and Ecology:
Occurring spontaneously on all islands and stable islets north of the Tropic of Capricorn as a strand plant or in coastal forest, from sea-level to about 600m, preferring well drained soils and frequently in pure sand within metres of the ocean. The importance of this palm in traditional Melanesian life is unsurpassed, providing food, drink, building material, household utensils, etc. Since colonisation of the region by European countries, the coconut has become an important economic crop from which is primarily derived coconut oil for use in the manufacture of soaps, perfumes, etc.

Description:
A tall palm to 30m with a curved trunk to 35cm in diameter, with prominent leaf scars. Roots are frequently exposed through erosion of the surrounding soil or sand. Leaves, about 15, are held erect to horizontal, becoming below the horizontal with age. Leafbases are clasping, not forming a crownshaft, fibrous based, cleanly abscising, the petiole short or long, stout, yellowish green with deciduous tomentum, the rachis to about 5.5m similarly covered. Leaflets, about 100 per side, are acuminate, closely

placed, light glossy green above, lighter below, with a prominent mid rib above with small ramenta on the mid rib below.

The interfoliar inflorescence, once branched, is at first held erect, becoming semi- pendulous in fruit, about 2m long. The prophyll is tubular, woody, persistent, the peduncular bract greatly exerted, plicate, woody, splitting below, with deciduous tomentum. The peduncle is stout, the rachis of about equal length, the rachillae spirally arranged, subtended by inconspicuous bracts. Triads are borne basally on the rachilla, paired or solitary male flowers distally. Male flowers are about 1.5cm long, yellow, sepals much shorter than the thick acute, valvate petals, stamens 6, the pistillode short with pointed lobes bearing nectar glands, the female flowers about 3cm long with three short triangular stigmas. The fruit is large, ellipsoidal to broadly ovoid, sometimes three angled, to 30cm long, green to yellow to ivory/grey at maturity, stigmatic remains apical. The epicarp is smooth, mesocarp very thick and fibrous, dry, endocarp thick and woody, more or less spherical, with three basal pores each with an operculum. The seed is large with a central cavity partially filled with fluid, the embryo basal.

DISTRIBUTION

Distribution of Genera in Nearby Areas:

As a preliminary to comparison of the genera which occur in the South- West Pacific, it may be relevant to review those which occur nearby in the Solomon Islands and Australia and not in the study area.

Of the 17 genera which are indigenous to the Solomon Islands, none of which is endemic, only seven are shared with either Fiji, Vanuatu or New Caledonia, whilst none is to be found on the islands at the higher latitudes. The shared genera include:

Licuala
Metroxylon
Calamus
Gulubia
Clinostigma
Physokentia
Cocos

The other genera which occur in the Solomon Islands include:

Livistona
Caryota
Nypa
Cyrtostachys
Ptychosperma
Drymophloeus
Areca
Heterospathe
Actinorhytis
Rhopaloblaste

Of the 19 indigenous Australian genera, six of which are endemic, four are shared with the South-West Pacific (Study Area). These include:

Licuala
Calamus
Gulubia
Cocos

The other genera which occur in Australia include:

Corypha
Livistona
Arenga
Caryota
Oraniopsis
Nypa
Archontophoenix
Linospadix
Laccospadix
Ptychosperma
Carpentaria
Normanbya
Wodyetia
Gronophyllum
Hydriastele

Genera in the nearby Islands to the East:

Due to the paucity of recent taxonomic activity, the identity of genera in nearby islands (Samoa, Tahiti, etc.) to the east is confused. An unclear determination of *Drymophloeus* (syn. *Solfia*) is listed (Langlois 1976) in Samoa to the north-east of Fiji, although not verified in recent authoritative literature (Uhl & Dransfield 1987). The early taxonomy of *Drymophloeus, Solfia* and *Balaka* in this region is not reliable. The palms of Samoa and Tahiti need revision to clarify this situation.

Clinostigma and *Metroxylon* occur in Samoa, and *Pritchardia* in Tonga.

Distribution of genera in the South-West Pacific:

The genera may be divided into four groups based upon their present distribution.

GROUP I

Those genera which are more diverse elsewhere but have species in the South-West Pacific:

Licuala
Pritchardia
Metroxylon
Calamus
Gulubia
Pelagodoxa
Clinostigma
Cocos

GROUP II.

Those genera which are most diverse in the South-West Pacific but have species elsewhere:

Veitchia
Balaka
Physokentia

GROUP III.

Those genera which are endemic to the South-West Pacific occurring on more than one island:

Rhopalostylis
Cyphophoenix
Cyphosperma

GROUP IV.

Those genera which are endemic to the South-West Pacific but confined to only one island:

Pritchardiopsis
Chambeyronia
Hedyscepe
Kentiopsis
Mackeea
Actinokentia
Howea
Neoveitchia
Brongniartikentia
Lepidorrachis
Clinosperma
Cyphokentia
Moratia
Alsmithia
Lavoixia
Carpoxylon
Alloschmidia
Campecarpus
Basselinia
Veillonia
Burretiokentia
Goniocladus

Endemism in the South-West Pacific:

Of the 36 indigenous genera, 25 are endemic. The majority, 16 genera, are in New Caledonia and of these nine are monotypic. These endemics are representative of subtribes Livistoninae, Archontophoenicinae and Iguanurinae. This diversity of palm genera on a single isolated island (except *Cyphophoenix* which has a species on Lifou, *Cyphosperma* which has species in Vanuatu and Fiji and *Cocos)* is unusual and unprecedented elsewhere in the world, perhaps for Madagascar.

New Caledonia has undergone a combination of conditions and events which has enhanced the retention and uninterrupted evolution of much of its archaic flora into the present. The basis of such includes an extant floristic source dating from at least the Mid Cretaceous (100 million years ago), and a coincidence of trans-latitudinal drift and uplift (in the Late Cainozoic, 2-3 million years ago) which prevented inundation and glacial devastation preserving much of the upland flora (Singh 1982). Thus the maintenance of optimum habitats and a diverse captive palm flora, though unsubstantiated in fossil records, has allowed evolution to proceed unhindered and develop and preserve the generic diversity.

Generic endemism elsewhere in the South-West Pacific is limited, being restricted to the volcanic islands, generally older, such as Viti Levu of Fiji which has two endemic genera. Firstly is *Neoveitchia* (monotypic) which has its closest, though tenuous, affinity in the Iguanurinae of New Caledonia and the Solomons, whilst secondly there is *Goniocladus,* an incompletely known taxon which has suspected affinity with *Physokentia. Alsmithia* is endemic to Taveuni.

Other endemic genera which do have obvious affinities include *Hedyscepe, Howea* and *Lepidorrachis* from Lord Howe Island, *Rhopalostylis* from New Zealand, Norfolk and Raoul Islands and *Carpoxylon* from Aneityum, in southern Vanuatu.

Hedyscepe and *Rhopalostylis* are closely related to each other and to *Archontophoenix* of north-eastern and eastern Australia, all of subtribe Archontophoenicinae; *Howea,* of Linospadicinae has an affinity with *Laccospadix* and *Linospadix* of eastern Australia whilst *Lepidorrachis* and *Carpoxylon* have an affinity with *Clinostigma* though with different cladistic references. *Rhopalostylis,* with two species and one subspecies and the monotypic *Carpoxylon* are found on recently formed volcanic islands.

Specific endemism is high, a natural consequence of the isolation of the archipelagos, island groups and islands within the study area. Apart from the pantropical *Cocos nucifera,* the extraterritorial species are to be found in New Guinea, Solomon Islands and Western Samoa. Most occur in the Santa Cruz Group of the Solomon Islands, about 320 kms. north of the Torres Group of Vanuatu. The Santa Cruz Group are geologically merged with the Solomon Islands and are considerably older than the islands of northern Vanuatu. The species shared with the Santa Cruz Group include *Licuala grandis, Gulubia cylindrocarpa* and *Metroxylon warburgii,* which is also shared with Western Samoa, whereas *Metroxylon salomonense* occurs in New Guinea, Santa Cruz Group and the Solomon Islands. It is suggested that the shared species occurring in Vanuatu represent the limits of the migrational radii of genera which are more diverse in areas to the north and north-west.

ORIGINS OF THE PALM FLORA

Fossil Evidence in the Region:

Palm fossils in the region are uncommon but most are found in deposits of Gondwanic origin. Fossils of *Nypa* are the most conspicuous, the extant taxa little changed from fossil specimens, undoubtedly the result of the unique habitat which it occupies. Sedimentation, and therefore fossil predisposition, is a by- factor of this palm's ecological balance and survival. *Nypa* has been found in deposits from the Upper Cretaceous (70 million years ago) from near Perth (Churchill 1973), Lower Eocene (53 million years ago) from Bass Strait (Partridge 1976) and New Zealand (Fleming) and from Eocene deposits from Queensland (Heckel 1972). Pollen referable to *Arecipites,* a palm prototype, appears in the North Maslin Sands, South Australia (Lange 1982) and date from Upper Eocene (40 million years ago) rocks.

Fossils of *Cocos* have been found in New Zealand. Fruits of *Cocos zeylanica* Berry and *Cocos* pollen appear through the Mid Oligocene (30 million years ago) to the Mid Miocene (20 million years ago). A record of *Rhopalostylis* pollen is reported from the Eocene (40 million years ago) (Fleming 1979) whilst an undated collection of fossil leaf impressions (Oliver 1911) is reported from the Meyer Islets, two small islands 1.5 kms north-east of Raoul Island. The Meyers are comparable in age to Raoul Island, from the Late Pliocene (2 million years ago).

Cocos zealanica. Endocarp (two views)(natural size)(from Fleming 1979)

Apart from *Rhopalostylis* in New Zealand and on Raoul Island, little similarity exists between the distribution of fossil species and extant taxa in the region.

Origins from a Subtribal Overview:

Before the commencement of the breakup of Gondwana (135 million years ago), the angiosperms (flowering plants) had been evolving. Fossil evidence indicates they were diverse and common in the western parts of Gondwana, now relative to present-day northern South America. They radiated rapidly to other parts of Gondwana as well as into the Northern Hemisphere land-mass.

The first palm fossils appear in the Upper Cretaceous (84 million years ago) in deposits from areas which had a warm moist climate with a temperature above 20°C. Palm fossils from cooler areas appear later. By the Mid Paleocene (60 million years ago), eastern Gondwana had achieved isolation through continental rift and separation. The vegetational captives, without input from outside genetic material, commenced evolving into their unique forms. This continent, now relative to present-day Australia, had progressively become arid as it drifted northward, the once dominant moist forests and optimum palm habitats retreating to areas in the north and east of the land-mass. These areas, now relative to the Top End of the Northern Territory and north-eastern Queensland, have been at times separated by advancement of the ocean across present-day western Queensland and central Australia.

New Caledonia, also in isolation from the Early Paleocene, has maintained much of its captive Gondwanan flora including the palms. Much of the extant archaic New Caledonian flora resembles that which was to be found in eastern Australia in the Late Cretaceous (Singh 1982). The New Guinea Region, which was to play an important role in migration and dispersal of much of the palm flora in the South-West Pacific, was at this time, a low extension on the northern edge of the continent and still remote from any South-East Asian or Malesian influence.

Colonisation of the Non-Continental Islands:

During the Early Paleocene (60 million years ago), the ensuing intermittent vegetational colonisation of the South-West Pacific commenced. Of the volcanically formed islands, Fiji was the first to emerge and thus receive vegetational material from nearby Gondwanic areas. It is possible that the first palms to reach Fiji were prototype Iguanurinae from New Caledonia. It was not until 30 million years later in the Early Oligocene that the Solomon Islands commenced to form, thus presenting an archipelic bridge which allowed a greater diversity of elements of the New Guinea flora and less directly, of the Australian flora to enter the South- West Pacific. It is possible that ancestral Ptychospermatinae may have gained access from this time. Upon examination of the extant Australian genera of this subtribe, it is suggested that this subtribe is of northern Gondwanan origin as demonstrated by the diversity of the indigenous Australian genera, some having become semi-arid adaptive, for example *Wodyetia* and *Carpentaria,* whilst others have remained inhabitants of rainforests and moist areas.

This adaptation of rainforest preferential taxa to drier conditions, as revealed explicitly in the fossil record, is a feature of other Australian families and could indicate by association, a long-time presence in northern Australia of Ptychospermatinae ancestors. It is suggested (Tucker 1988) that at least some taxa within Ptychospermatinae which are presently in the north-eastern Australian rainforests (*Ptychosperma macarthurii*) have arrived through recent direct southward migration from New Guinea with the Torresian or Nutmeg Pidgeon as vector. The New Guinea and Solomon Island's diversity of Ptychospermatinae may represent a response initiated after migration from the south into more amenable habitats and climates.

Ten million years after the Solomons emerged in the Late Miocene (20 million years ago), so did the oldest islands of Vanuatu. Migration of the vegetation from the Solomons, Fiji and New Caledonia would have been possible. Most prominent of the extant Vanuatu palms are the Ptychospermatinae, which are of Fijian affinity and Iguanurinae of New Caledonian affinity. Peripheral representatives of Calaminae, Metroxylinae, Arecinae and Livistoninae have also entered the archipelago, possibly at a similar time as they did into Fiji, from the north.

Representatives of the widespread Livistoninae have become established in Vanuatu and Fiji, the former with *Licuala* and the latter with *Pritchardia,* but both by migrational progress. Livistoninae is scantly represented in the South- West Pacific considering the diversity in New Guinea and Australia. It appears to have a predilection for continental non-volcanic areas, at least in the Old World regions of Gondwanic origin. The occurrence of *Pritchardiopsis* in New Caledonia represents a retention of the archaic Gondwanan flora.

Introduction of subtribes Arecinae, Calaminae and Metroxylinae could have commenced once the northward drifting abutment of New Guinea had approached the islands of Malesia in the Late Miocene (12 million years ago), if it is supposed that these subtribes were established in Malesia and not already present in the Gondwanan region. A possible suggestion is that ancestors of these subtribes were carried to the Northern Hemisphere by the northward rafting Indian subcontinent, after its detachment from Gondwana. Arecinae and Calaminae are presently most diverse in South-East Asia, whereas Metroxylinae straddles the convergence of the Gondwanan and Northern Hemisphere land masses.

The migrational problems of *Metroxylon* have been addressed previously (Johns & Hay 1984). The large seeds fail to survive even short periods of inundation in sea-water and neither do birds nor bats eat the fruits. Interestingly, it is reported (Tucker 1988) that whole *Metroxylon sagu* trunks are not uncommonly washed ashore in north Queensland. For this genus, distribution through whole plants being dislodged and carried on the ocean's currents, should be considered.

Anomalous is the presence of the Arecinae genera *Gronophyllum* and *Hydriastele* in northern Australia, the former fire and semi-arid adaptive and indicative of a long-time presence whilst the latter is a strict inhabitant of moist areas also being conspecific in eastern Australia.

Nevertheless, these three subtribes are represented by very few taxa in the South- West Pacific, not far evolved from those nearby in the Solomons and New Guinea.

Meanwhile to the south in the Mid Oligocene (25 million years ago), Lord Howe Island emerged through volcanic activity. Ancestral Linospadicinae, Archontophoenicinae and Iguanurinae would have progressively arrived from nearby Gondwanic areas by way of island hopping or direct migration.

Contradictory migrational theories have been presented by various authors for vegetational establishment on Norfolk and Raoul Islands. The former emerged in the Mid Pliocene (2.5 million years ago) whilst Raoul emerged .5 million years later in the Late Pliocene. One suggestion is that *Rhopalostylis,* with a long- time presence in New Zealand since at least the Mid Eocene (40 million years ago) as a relic of Gondwanan flora, firstly migrated to Norfolk to the north-west and from there eastward to Raoul. It is also suggested (Fleming 1979) that *Rhopalostylis* is of Malayo-Pacific origin (and not of archaic Gondwanan flora) arriving during the Tertiary, whilst it is suggested (Sykes 1977) that *Rhopalostylis* on Raoul may be the result of direct migration from New Zealand and not Norfolk due to a predominantly north-west ocean current. This suggestion is supported by the number of other taxa which are present in both north New Zealand and on Raoul but not on Norfolk. Nevertheless, the present-day *Rhopalostylis* from Norfolk and Raoul are more similar to each other than to the taxon in New Zealand.

It is suggested that the occurrence of Butiinae (*Cocos nucifera*) in the South-West Pacific is recent, despite the recording of fossil material of a now extinct species, *C. zealanica* Berry in New Zealand. It is evident that all *Cocos* in the South-West Pacific are of cultivated origin, or their progeny, introduced by man although it is reported (Tucker 1988) that populations which may represent a truly indigenous type of *Cocos nucifera* are found on the coasts of north-eastern Australia, with a particularly distinct taxon on the west coast of Cape York Peninsula. They have small spindle shaped fruits. Eastward migration into the South-West Pacific from these northern Australian areas seems implausible when ocean currents are studied.

In summary, it is probable that all present-day palms in the South-West Pacific are of Gondwanic origin. Their presence in the region has been the result of retention of the Gondwanan flora, as with Livistoninae, Archontophoenicinae and Iguanurinae or by migration, both direct and indirect, from a number of possible sources. Of possible direct migration from an eastern Australian area are Linospadicinae and Archontophoenicinae whilst Ptychospermatinae and Livistoninae display indirect migration. Indirect migration via an Asian source is possible for Metroxylinae, Calaminae, Arecinae and a secondary branch of Livistoninae. Butiinae has been introduced by man.

COMMON AND VERNACULAR NAMES OF THE PALMS OF THE SOUTH-WEST PACIFIC

FIJI :

Balaka seemannii, B.longirostris; mbalaka, niuniu, niu mbalaka, nesi.
B.macrocarpa; niuniu.
B.pauciflora; black bamboo.
Calamus vitiensis; watem buretathi, mganua, ngganua.
Clinostigma exorrhizum; niu niu, niu sawa.
Cocos nucifera; niu, coconut, niu ndina, niu yabia (weeping coconut).
Cyphosperma tanga; tanga.
Goniocladus petiolatus; natana.
Metroxylon vitiense; sago, songa, niu soria, sogo.
Neoveitchia storckii; vuleito.
Physokentia thurstonii; niu niu.
P.rosea; tangandanu.
Pritchardia thurstonii; masai, niu sawa, viu.
P.pacifica (not included in this work); masei, niumasei, niusakiki, sakiki, viu, Fiji fan palm, saii.
Veitchia joannis; niu sawa, niu niu, sangewa.
V. simulans; niu sawa.
V. pedionoma, V. petiolata and *V. sessilifolia;* niu niu.
V. vitiensis; nesi, kaivatu.

VANUATU:

Calamus sp.; lelaoros, rotin, kaula, kaekaula, garu fwala, nowol, aluahne, garwol.
Carpoxylon macrospermum; ninuvisu.
Clinostigma harlandii; ngami igh, kolkol, olo olo, patub.
Cocos nucifera; niu, coconut, coconus, kavra, kopra, navar, samsam, kul, nium weto, liul.
Gulubia cylindrocarpa; motoval, niulip.
Licuala grandis; umbrella leaf, pune pele, tabataba, sonje, lali.
Metroxylon warburgii; netato, tsuku,
natangura, natalawa, enkul, natakura, tangula, wataghor, sokora, notah.
Physokentia tete; tete, vhe.
Veitchia spp.; nal, kulkul, nawabuk, ngup, huup, nival, nokimpat, wael coconus, wogha mi ir, wahuru, borkul, totougatutu.
V.arecina; napupwa, napupuwa.
V.metiti; metiti.
V.winin; winin.

LORD HOWE ISLAND:

Hedyscepe canterburyana; big mountain palm.
Howea belmoreana; curly palm.
H.forsteriana; kentia palm, thatch palm.
Lepidorrachis mooreana; little mountain palm.

NEW CALEDONIA :

Cyphokentia macrostachya; white palm, palmier blanc.
Cyphophoenix nucele; nucele.

NORFOLK ISLAND:

Rhopalostylis baueri var. *baueri;* Norfolk Island nikau.

RAOUL ISLAND:

Rhopalostylis baueri var. *cheesemanii;* Raoul Island nikau.

NEW ZEALAND:

Rhopalostylis sapida; nikau.

PARC TERRITORIAL DE LA
RIVIERE BLEUE
LE KAORI GÉANT
LES CASCADES
MONTAGNE DES SOURCES
MOIS DE MAI
LAC DE YATÉ
RIVIERE BLANCHE
LES ELECTRIQUES
LE BON SECOURS
LE PONT GERMAIN
LE PONT PERIGNON
LIMITES DU PARC TERRITORIAL
ROUTE CARROSSABLE
PARCOURS PEDESTRES
REFUGES TABLES
CASCADES
BAIN
R. BLEUE
R. BLANCHE
LAC DE YATE
YATÉ 38km
NOUMEA 45km
RENÉ RENNETEAU
1985

MORPHOLOGY, DISTRIBUTION AND ECOLOGY OF PALMS IN NEW CALEDONIA

T. Jaffre & J.-M. Veillon
ORSTOM, Noumea
Translated by Marie Trefeu, edited by J.L. Dowe

Introduction:

The palms of New Caledonia belong to two subfamilies, the Coryphoideae represented by the genus *Pritchardiopsis* with palmate leaves and bisexual flowers, and the Arecoideae with sixteen genera totalling thirty one species with pinnate leaves and unisexual flowers in triads. Among those last named, the genus *Basselinia* with eleven species in two sections is the most diverse. Only the genus *Cyphosperma,* present in Fiji and Vanuatu, has an extra-territorial distribution.

This article's aim is to present the morphological and architectural characteristics of palms in New Caledonia, to define their distribution regulated by ecological factors in their environment and also their localities in New Caledonia.

Morphological Architecture:

All the palms in New Caledonia are phanerophytes, not one species is climbing nor do any have spines.

At maturity, the majority are 10 to 18 metres in height. Among the smaller species there are *Basselinia deplanchei* (4-5m), *B. gracilis* (2-8m), *B.iterata* (7m), *B.porphyrea* (5m), *B.vestita* (.5-5m), *Brongniartikentia lanuginosa* (1-6m) and *B.vaginata* (1-6m). In taller species there are *Basselinia tomentosa* (20m), *Burretiokentia vieillardii* (18-20m), *Kentiopsis oliviformis* (30m), *Mackeea magnifica* (20-25m) and *Moratia cerifera* (20m).

All the species are unbranched with slow growth and smoothish trunks with more or less prominent annular rings. One-third of the species have prominent annular rings, remarkable in *Basselinia humboldtiana, B.sordida, Brongniartikentia lanuginosa, Burretiokentia vieillardii* and *Campecarpus fulcitus.*

Most species have a swollen trunk at the base, forming in some cases *(Cyphokentia macrostachya, Lavoixia macrocarpa)* circular disc at ground level. One-third of the species have aerial roots and three of them *(Alloschmidia glabrata, Burretiokentia vieillardii, Campecarpus fulcitus)* possess stilt roots arranged in several cones.

They all have a spiral phyllotaxy and the number of functional leaves varies from four *(Actinokentia divaricata)* to about twenty *(Cyphosperma balansae).* All the *Basselinia* of section Taloua have about ten.The inflorescences are lateral, situated under the crownshaft, except in the two species of *Brongniartikentia,* in *Cyphosperma balansae* and *Pritchardiopsis jeanneneyi* where they are among the leaves. The number of inflorescences varies, but is most often two, although more than ten can be attained by *Alloschmidia glabrata* and *Cyphosperma balansae,* however, in that species and in *Brongniartikentia vaginata* the inflorescences are situated at the extremity of a long peduncle which can attain 1.5 metres.

The architecture of palms in New Caledonia (Veillon 1976) corresponds to two models defined by Halle' et al (1978), the model of Corner for species with solitary trunks, the model of Tomlinson for the four caespitose species; *Basselinia deplanchei, B.gracilis, B.vestita* and *B.pancheri.* The case of *B.pancheri* is very interesting to note. This palm most often presents the model of Corner but occasionally (Moore & Uhl 1984) it branches at the base of the trunk which makes it conform to the

Figure 74: Opposite previous page. Upper left. *Basselinia pancheri* in habitat, slopes above Lake Yate, southern New Caledonia (AD)
Upper right. *Basselinia deplanchei* in habitat (O)
Lower. The view downstream from Pont Germain toward the entrance of Lake Yate, Riviere Bleue, New Caledonia (AD)

Figure 75: Previous page. Upper left. Infructescence with mature fruit of *Cyphokentia macrostachya,* Riviere Bleue basin(AD)
Upper right. Crown of *Cyphokentia macrostachya* (AD)
Lower. Location board of the Riviere Bleue Terratorial Park (AD).

model of Tomlinson. This example illustrates the possible transition between the two close architectural models within the same genus. Outside the forest environment, the adaption of the caespitose species in full light is marked not only for an important reduction in size or a lesser number of leaves which become tougher, but also by a multiplication of stems often reduced to simple clusters at the base (10-12 stems) for *Basselinia deplanchei* in the "maquis" of Haute Tontouta and Ouenghi. Generally, morphological transformations concerning essentially the reduction of the size of the trunk, the leaves, the spathes and of the inflorescences are observed in a good number of species when they are in an overexposed environment.

Distribution and Ecology:

The distribution of palms in New Caledonia has been approached by different authors (Schmid 1973, 1974: Moore & Uhl 1984: MacKee et al 1985).

We can recognise five different phytogeographic sectors including within them areas of dense humid forest and the scrub derived from it.

The south is constituted essentially of ultrabasic rocks, the north-east of micaschists and glaucophanites and the others of ultrabasic and acid rocks (schists and micaschists).

The distribution of the species, the nature of the geological substratum and the edaphic conditions are given in Table I.

Geographic and Geological Distribution:

The southern sector, essentially constituted of ultrabasic rocks (mainly peridotites), has fourteen species. Six of them belong to five genera *(Actinokentia huerlimannii, Basselinia humboldtiana, B.porphyrea, Brongniartikentia vaginata, Campecarpus fulcitus, Pritchardiopsis jeanneneyi)* are strictly associated with them. We also find *Burretiokentia vieillardii* and *Basselinia gracilis* from north to south on La Grande-Terre on various substrata. *Clinosperma bracteale* and *Cyphokentia macrostachya* are present in the central sector on acid rocks. Three species, *Actinokentia divaricata, Basselinia deplanchei* and *B. pancheri,* are localised in ultrabasic regions, the first two being distributed in the south, central and west, the third in the south and north-east.

Nine species are found in the east-central sector but only one species, *Basselinia tomentosa,* reported in only one location on schists (Mt. Nakada), is endemic to that region. *Basselinia sordida* is associated with ultrabasic rocks, the other six are indifferent to the substratum.

The central-west sector has 13 species. Only *Basselinia vestita,* known in a single locality, Le Me' Ori on peridotites, is endemic to that sector. four other species, including *Actinokentia divaricata* and *Basselinia deplanchei* are also found in the south sector. *Basselinia deplanchei,* with a large distribution, is always associated with ultrabasic rocks. Two species, *Kentiopsis oliviformis,* which is exclusive to the central-west sector and *Basselinia velutina,* the only species having a disjunct distribution (present at Dogny in the central-west sector and at Mt.Panie' in the north-east sector) are associated with acid rocks. Six species are indifferent to the substrata and all except *Cyphosperma balansae* are present in the south sector.

Of the species found in the north-east sector, only *Burretiokentia vieillardii* (the only species represented in the five phytogeographic regions) is indifferent to the substratum. *Basselinia pancheri* is exclusive to the ultrabasic rocks in the south and central sectors and *Burretiokentia hapala* is associated with limestone rocks (but more frequently on acid rocks) in the north-east. Sixteen species have been found in the north-east sector. Ten of them, with five being monotypic genera *(Alloschmidia, Lavoixia, Mackeea, Moratia* and *Veillonia)* and five belonging to *Basselinia* (two species), *Chambeyronia, Brongniartikentia* and *Cyphophoenix,* are endemic to that sector. Of the last two genera, *Brongniartikentia lanuginosa* is found on Mt. Panie' and *B. vaginata* on ultrabasic rocks in the south

Figure 76: Following page, Upper left. *Basselinia gracilis* in habitat (AD)
Upper left. Leaves of *Basselinia pancheri* (AD)
Lower. Mature infructescence of *Basselinia pancheri;* note the kidney shaped fruit (AD)

Figure 77: Opposite following page. Upper left. *Cyphophoenix nucele* on Lifou Island, Loyalty Islands, New Caledonia (O)
Upper right. *Lavoixia macrocarpa* in habitat at 500m asl on Mt. Panie', north-eastern New Caledonia (O)
Lower. Fruit of *Pritchardiopsis jeanneneyi* (O)

Species	S	C.E	C.W	N.W	N.E	Habitat	Acid or Calcium Rocks	Peridotes	Altitude (m)
Actinokentia divaricata	X		X			F U		e a	200-700
Actinokentia huerlimannii	X					F U		e r	850-1000
Alloschmidia glabrata					X	F A			10-1000
Basselinia deplanchei	X		X			M U		e f r a	500-1500
Basselinia favieri					X	F A	f		300-500
Basselinia gracilis	X	X	X		X	M I	e f	e f r	10-1600
Basselinia humboldtiana	X					F U		e r	800-1000
Basselinia iterata					X	F A	e		1000
Basselinia pancheri	X		X	X		M U		e f r	200-1200
Basselinia porphyrea	X					F U		e r	900-1000
Basselinia sordida		X	X			F U		r	100-1400
Basselinia tomentosa		X				F A	e		1000-1100
Basselinia velutina			X		X	F A	e f		400-1600
Basselinia vestita			X			F U		e	900-1000
Brongniartikentia lanuginosa					X	F A	e f		700-1250
Brongniartikentia vaginata	X					F U		e f a	200-1000
Burretiokentia hapala				X	X	F A C	e c		50-400
Burretiokentia vieillardii	X	X	X	X	X	F I	e f	e f r	900-1200
Campecarpus fulcitus	X					F U		e f	20-800
Chambeyronia lepidota					X	F A	e f		300-1400
Chambeyronia macrocarpa	X	X	X		X	F I	f	f a	10-800
Clinosperma bracteale	X	X	X			F I	e f	e f r	40-1200
Cyphokentia macrostachya	X	X	X			F I	f	e f a	150-900
Cyphophoenix elegans					X	F A	f		20-550
Cyphophoenix nucele						F C	c		60
Cyphosperma balansae		X	X		X	F I	f	f	10-950
Kentiopsis oliviformis		X	X			F A	e f		30-300
Lavoixia macrocarpa					X	F A	f		500
Mackeea magnifica					X	F A	f		500-700
Moratia cerifera					X	F A	f		10-800
Pritchardiopsis jeanneneyi	X					F U		e	200
Veillonia alba					X	F A	f		10-600

Legend for Table I:

Geographic Separation: S-South, C.E.-Central East, C.W.-Central West, N.W.-North West, N.E.-North East

Plant Formation, Geological Substrata:

FU - forest on ultrabasic rocks
FA - forest on acid rocks
MU - forest and marquis on ultrabasic rocks
MI - forest on acid rocks, forest and marquis on ultrabasic rocks
FI - forest on acid and ultrabasic rocks
FC - forest on limestone
FAC - forest on acid rocks and limestone

Soils:

On Acid Rocks or Limestone

e - developed soils on schists, micaschists and glaucophanites
f - ferrallitic soils on schists, micaschists and glaucophanites
c - brown soils on limestone

On Ultrabasic Rocks

e - developed soils on peridotites
f - humiferous ferrallitic, ferritic soils on peridotites
r - juxtaposition of developed soils and organic rankers on peridotites
a - serpentine alluviums

Figure 78: Table 1. Distribution of Species, Nature of Geological Substratum and Edaphic Conditions.

whilst *Cyphophoenix elegans* on the Panie' massif and *C.nucele* on limestone on Lifou. They are examples of genera adapting to different soil conditions.

Cyphophoenix nucele, located at only one place at Lifou, has thc peculiarity of being the only palm endemic to the Loyalty Islands. It is necessary to acknowledge the absence of palms from the Isle of Pines.

The number of species on different categories of geological substrata shows that thirteen species (40.6% of the total) are associated with schists and micaschists. Eleven species (34.4%) are associated with mining terrains. Six species (18.75%) are found either on acid rocks or on ultrabasic rocks. Only one species is associated with limestone and another is found on both limestone and acid rocks.

The importance of palms in the south sector (nine genera and fourteen species, of which ten are endemic), demonstrates the bipolarity of their distribution. However, the north-east sector, richer, despite an area about seven times inferior to the south sector, appears to be the favorite sector of the family.

In respect to palms being exclusively restricted to forests, a thesis developed by Morat et al 1984, concludes that the moist forests, and particularly those of the north-east on micaschists, could have had a preponderant role in the conservation and diversification of the flora in the forests of New Caledonia. This role, conservator and diversificator, is also played but to a lesser degree, by the forests on ultrabasic rocks and particularly in the large south Massif which occupies one quarter of La Grande-Terre and culminates as does Mt.Panie' at more than 1,600m in altitude at Mt.Humboldt at 1,618m.

Edaphic Conditions:

It appears as indicated in Fig.I. that the palms of New Caledonia are generally associated with dense, humid forests.

Four species only occur also in "maquis" derived from forest on ultrabasic rocks and are found on seven main types of soil whose characteristics are mainly determined by the geological substratum. On schists, micaschists and glaucophanites they occur on or at the base of steep slopes of immature eroded soils or in flattish areas of slightly leached fersiallitic or ferrallitic soils.

The undeveloped humiferous soils of the metamorphic Mt. Panie' (Latham 1985), have a silty texture. They are rich in organic matter (5-10%), poor in nitrogen (0.4%), very acidic (pH 3.9 to 4) and very desaturated in bases. They are poor in phosphorous (0.3% of P_2O_5in the humifere horizon) but have a very high reserve of potassium (0.7% to 0.9% of K_2O).

The fersiallitic and ferrallitic soils are less desaturated than the previous ones, their pH often being near 5, but they are relatively poor in nutritional elements like phosphorus. They are distinguished, especially from underdeveloped soils, by a deeper and distinctly differentiated pedologic profile. They offer more favorable conditions for root development and alimentation.

On limestone, *Burretiokentia hapala* (only one location in this case) and *Cyphophoenix nucele* are found on brown soils in juxtaposition with fersiallitic soils for the first one and skeletic for the second. These neutral or slightly acid soils (Latham and Mercky 1983) are silty clay of lumpy structure, rich in organic matter (more than 20%) and in nitrogen (average 1%). Among the exchangeable bases, calcium dominates (more than 15 me'/100g). The reserves in phosphorus are very high (2%) on limestone at Lifou and very low (0.1% of P_2O_5) on limestone in the north-east sector. In general, these soils are more fertile than the soils on acid or ultrabasic rocks particularly at Lifou, because they are too shallow to favour a good water supply in dry periods.

On ultrabasic rocks, the four categories of soils named offer to the palms very special conditions of mineral nutrition because of their extreme poverty in every element except magnesium, sometimes in excess, and more or less high content in heavy metals (nickel, chromium, copper and manganese).

In every case, their total contents in calcium and potassium and in phosphorus are extremely low (in order of 0.01%). In the exchange complex the magnesium is the dominant element: from 5 to 10 me'/100g on underdeveloped soils in erosion, inferior from 2 me'/100g on ferrallitic soils and also on the organic rankers (3 to 3.5).

The non-developed soils on alluvial deposits, or on the slopes, have a very high content in nickel (from 0.5 to 1%). Although palms are frequently close to species rich in nickel (more than 1% of dry matter in

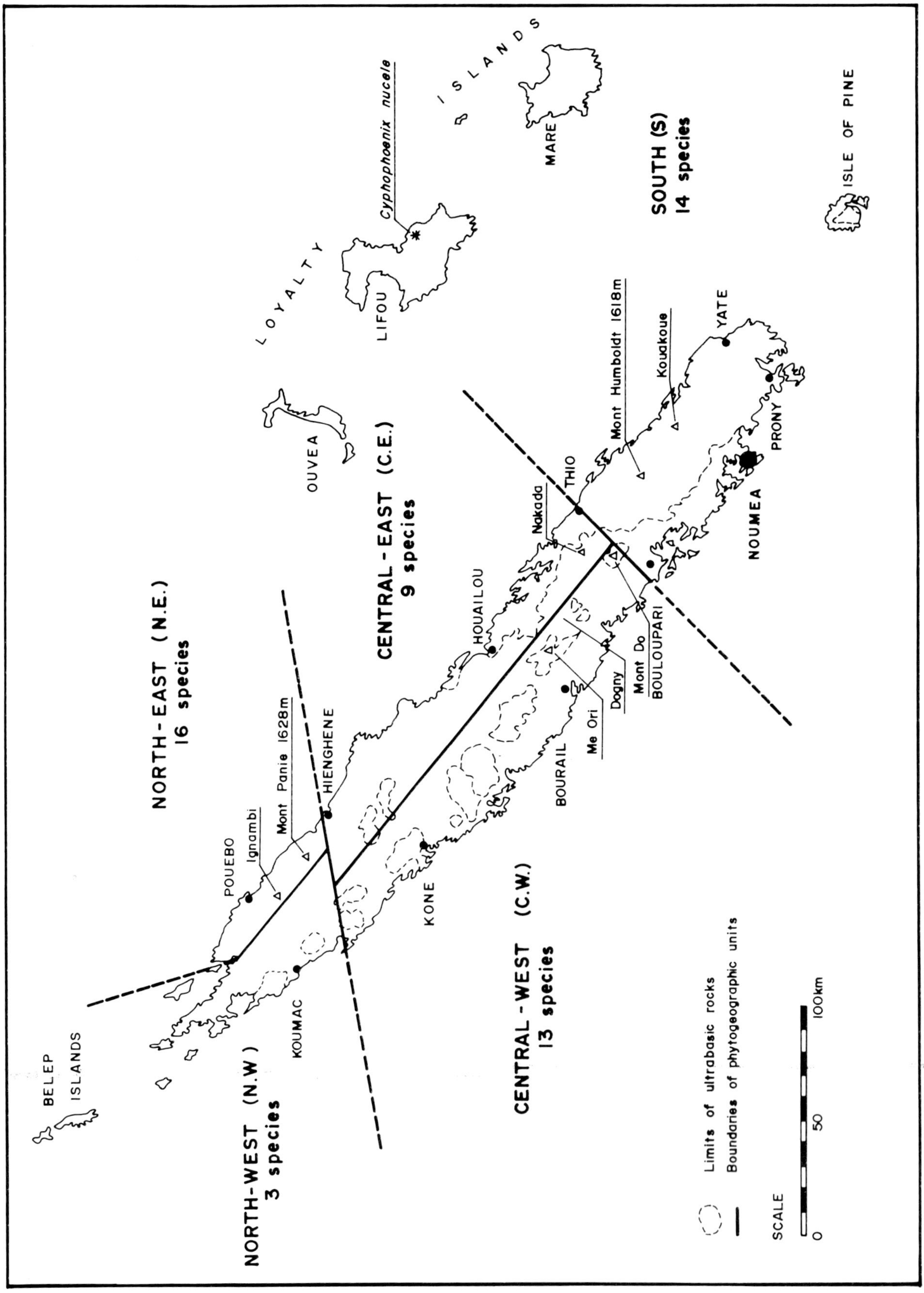

Figure 79: (Fig. 1.). Distribution of Palms in New Caledonia.

Sebertia acuminata, Psychotria douarrei and *Hybanthus austrocaledonicus),* none of them has a very high content of nickel. Palms associated with soils rich in nickel on ultrabasic rocks, like monocotyledons in general (Jaffre' 1980), would resist the toxicity of nickel by limiting the penetration of this element in their tissues. It appears, and this tends to be proven by the successful culture of many palms associated with mining terrains on substrata well balanced in nutritious elements, that species capable of enduring the conditions of mineral nutrition on ultrabasic rocks have no specialphysiological requirements in mineral nutrition.

On very deep alluvial soils, palms have to face a temporary hydromorphy of the soil. On ferrallitic soils and on gentle slopes, the root system development is impeded by the shallow rocky soil. It is the same on ferrallitic ferritic soils where *Basselinia pancheri* is found and in the organic rankers on which the humiferous soils can attain up to 1m in depth. This explains why the high altitude palms are found preferring organic rankers in juxtaposition with soils not affected by erosion.

Distribution in Relationship to Climatic, Microclimatic and Altitudinal Conditions:

All the palms in New Caledonia are found in dense humid forests (rainforest). Three species, *Basselinia deplanchei, B.gracilis* and *B.pancheri* are also found in caespitose form in the maquis derived from forest on ultrabasics, as observed particularly in the massifs of Kouakoue, Humboldt, Boulinda and Tchingou.

There are no palms under the rainfall level of 1,200mm per annum. As a result, they are absent from the sclerophyllous forests on the west coast which are associated with an annual rainfall not exceeding 1,100mm per annum. The palms of low altitude are more concentrated on the east coast (particularly at the base of Mt.Panie': Fig. 2.) and in the well watered south massif. On the west coast they are found only under 300m on river banks in gallery forest.

The altitudinal distribution of every species is given in Table I. Twenty-seven species (85% of the total collection) are present between 300m and 1,000m in altitudes which correspond with the dense humid evergreen forests of low and medium altitude (Morat et al 1981) which receive between 1,500mm and 3,500mm of rain per annum. Most species belong to the understory of the forest, however, five species, *Burretiokentia vieillardii, Cyphokentia macrostachya, Kentiopsis oliviformis, Lavoixia macrocarpa* and *Mackeea magnifica* may emerge from the forest top which in these forests is about 20 metres high.

A study of the density of the palms in rainforest, on either the slopes or on the alluvium, has been undertaken on the Riviere Bleue in the south massif. The results (Table II) indicate that *Actinokentia divaricata, Basselinia pancheri, Campecarpus fulcitus* and *Cyphokentia macrostachya* are found in the two types of station, whilst *Basselinia gracilis* and *Chambeyronia macrocarpa* exclusively on alluvium and *Brongniartikentia vaginata* exclusively on the slopes. They all have diameters less than 20cm and belong to the understorey, occupying the interstices between the tall trees.

	ALLUVIUM				SLOPES			
	No. of Palms per hectare	%	Surface area cm^2	%	No. of Palms per hectare	%	Surface area cm^2	%
2-5cm	27	0.84	475	1.37	186	3.18	2293	4.84
5-10cm	159	7.40	5891	6.82	517	19.49	20753	19.80
10-20cm	32	4.09	4112	3.44	223	19.61	27799	16.65
>2<20cm	229	3.86	10479	3.88	926	14.09	30845	13.75

Figure 80: Table II. Comparison of Palms in Numbers and Ground Surface Area Covered in Forest on Alluvium and on Slopes in the Riviere Bleue Basin (South Massif).

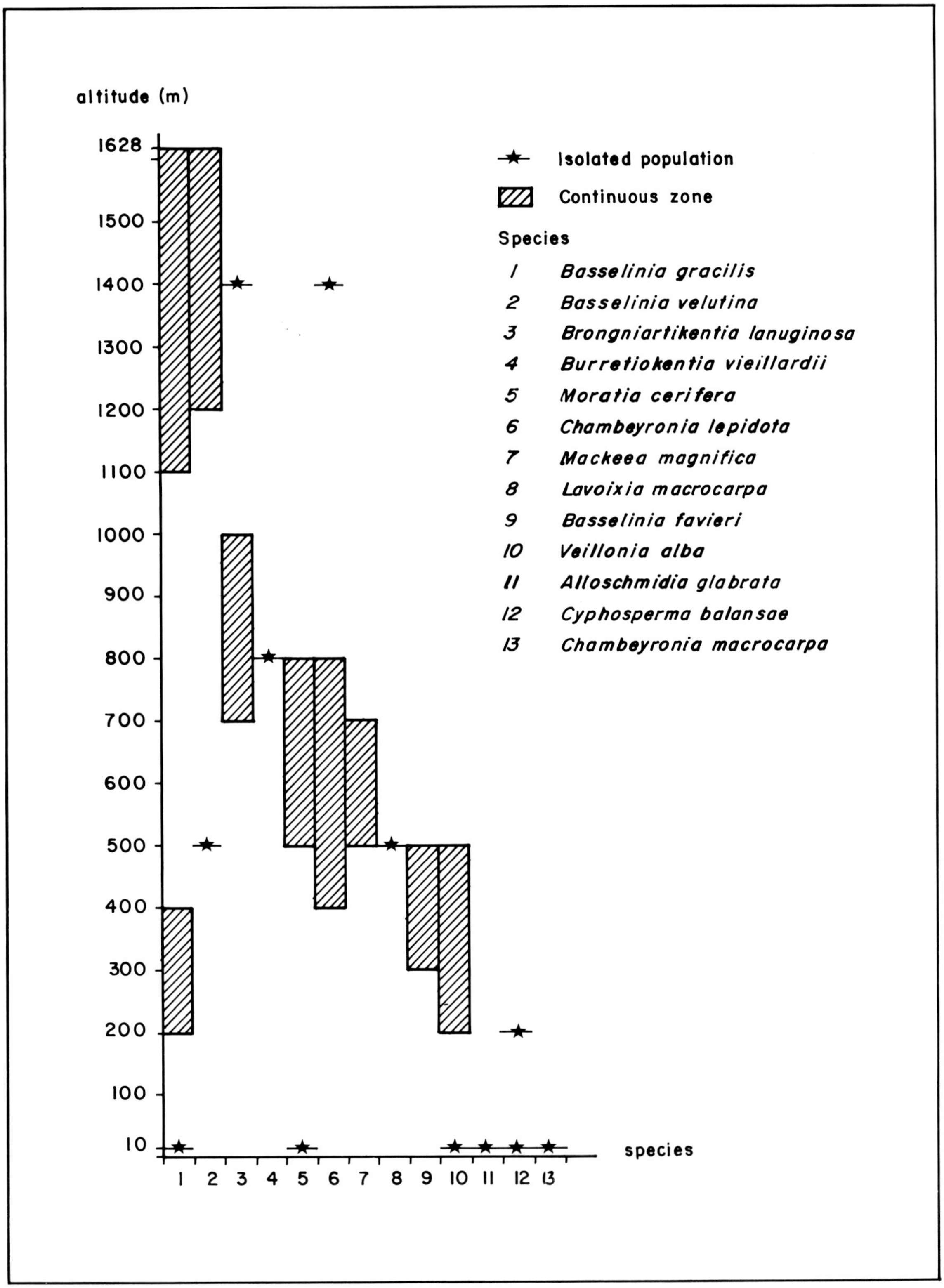

Figure 81: (Fig. 2). Altitudinal Distribution of Palms on Mt. Panie' between 10m (Creek Pouai) and 1628m (The Summit).

It also indicates that the palms are relatively more important on the slopes where they preferably occupy ancient chablis (naturally open areas), a phenomenon also observed in French Guyana by Granville (1978), than on the alluvium where the canopy is more level and closed.

In forests on alluvium where the diameter class is less than 20cm, the palms represent 3.86% of the tree and shrub population and 3.88% of the ground surface (ground surface: total of the trunk sections at 1.3m high). On the slopes these values are respectively at 14.09% and 13.75%.

It is in the range from 5cm to 10cm in diameter that palms are in the greatest number, with 159 per hectare on alluvium and 517 on the slopes. In the last instance, the importance of palms on ground surface increases between 10cm to 20cm in diameter.

In comparison to forest plants with a diameter greater than 2cm, the palms represent respectively 2.97% of population and 9.28% of ground surface on the slopes and 1.77% and 4.77% on flat surface.

Though below the values given by Kahn (1986) for the Amazonian forest in the Rio Negro Basin (2,122 palms per hectare and 32 species per hectare), the results show the importance of palms in dense humid forests (rainforests) at low to medium altitudes in New Caledonia which gives, especially on the slopes, a distinct appearance.

Only three species, *Basselinia iterata, B.sordida* and *B.tomentosa,* all of section Taloua, are strictly associated with the dense, humid evergreen forest at high altitude (above 1,000m), but fourteen others are also found in altitudes attaining 1,000m (Table I). This means that more than half of New Caledonian palms are found in forests frequently covered with clouds, receiving 3,000mm to 4,000mm of rain per annum and subjected to low temperatures, close to 0°C in the cold season (July- August).

Among the species associated with high altitude forests, only *Basselinia iterata* is exclusive to the undergrowth, the others frequently emerging from above the trees which are most often 15m in height.

Rare and Endangered Species:

Palms occupy a primordial position among the monocotyledons of the New Caledonia forests, as illustrated by their importance in certain biotopes and their richness in species. With 32 species, they constitute about 13.5% of the monocotyledon flora in the forests. This abundance cannot however hide the fact that many species are rare and even threatened with extinction.

A number of species are locally abundant, including *Basselinia humboldtiana, B.porphyrea* and *Actinokentia huerlimannii* of La Montagne des Sources and Mt.Humboldt between 800m and 1,000m and *Basselinia favieri* on Mt.Panie' between 300m and 500m.

Two species, *Basselinia iterata* known only at Mt. Ignambi and Mt. Colnett at 1,000m and *B. vestitia* of Me' Ori between 900m and 1,000m, should be considered endangered because of their restricted locations.

Cyphophoenix nucele, Lavoixia macrocarpa and *Pritchardiopsis jeanneneyi,* known only in one location with a very limited number of individuals, are rare species threatened with extinction. They deserve particular attention.

Cyphophoenix nucele:

This species, discovered by Daniker in 1925, has a population of about 100 individuals, concentrated in one location on madreporic limestone on the Isle of Lifou (Loyalty Islands). The population is situated in dense forest on raised coral terraces and is naturally protected from bushfire and destruction by man. The population consists of individuals of different ages which indicates a good reproduction of the species.

Lavoixia macrocarpa:

Discovered more than 20 years ago (1968) on Mt.Panie' by L.Lavoix, this species is known to have only five plants, (four adults) distributed on less than 0.5ha. Although the species bears abundant fruit, no germination has been observed in the vicinity of the trees. After many unsuccessful attempts to germinate seeds in the laboratory, the survival of this species seems to be threatened in the more or less long term. However, the search for other populations on nearby slopes and areas with difficult access merits to be pursued.

Pritchardiopsis jeanneneyi:

Collected for the first time by Jeanneney in 1893, this species has had, as emphasised by Morat, (1986) a history rich in changing fortunes. Although it was believed extinct at the beginning of this century, supposedly destroyed by the convicts from the penal settlement at Prony, it was found again in 1977 by a hunter, J.C. Lecren. He notified in 1980 R. Aymard who a few years earlier had participated in the fruitless search for this palm, organised by Lavoix. A specimen collected by these two was immediately authenticated by Dr. H.E. Moore Jr. who was present in Noumea at the time. The unique population of this species, spread on about 1 ha, is of three small populations totalling 30 plants with only one fertile adult of about 12m high. All the others have a size less than 4m. This disparity in the size of individuals could be the result of a vegetative dormancy or the consequence of an unreliable reproduction in the field. However, seed production has been regular since 1980 and numerous germinations have been obtained in the laboratory.

Lavoixia macrocarpa and *Pritchardiopsis jeanneneyi* are the beneficiaries of a careful surveillance by the Forests and Natural Patrimony Service of New Caledonia. Seeds of these two species and seedlings of *Pritchardiopsis jeanneneyi* have been sent to various botanical gardens in France and elsewhere in order to diversify the chances of reproduction and survival.

REFERENCES:

GRANVILLE (de) J.J. (1978) Les Monocotyledones dans la Foret Guyanaise. In Recherche sur la Flore et lat Vegetation Guyaniases. These de Doct. de'etat U.S.T.L. Montpellier 272pp

HALLE F., OLDEMAN R.A.A. and P.B. TOMLINSON (1978). Tropical Trees and Forest. An Architectural Analysis. Springer Verlag, Berlin, Heidelberg and New York.

JAFFRE T. (1980). Etude Ecologique du Peuplement Vegetal des sols. Derives de Roches Ultrabasiques en Nouvelle-Caledonia. Coll. Trav. et Doc. de l'ORSTOM numero 124. 274pp.

KAHN F. (1986). Adaption en Foret Tropicals Humide: Le Cas des Palmiers Amazoniens. Naturalia Monspeliensia. Colloque International sur L'Arbre. pp. 171-189.

LATHAM M. (1985). Carte des Sols au 1:50,000 Pam-Ouegoa. Notice explicative numero 101. ORSTOM Paris.

LATHAM M. et MERCKY P. (1983). Etude des Sols des Iles Loyaute. Carte pedologique au 1:200,000. Notice explicative numero 99, ORSTOM Paris.

MACKEE H.S.; MORAT Ph. and VEILLON J.-M. (1985) Palms in New Caledonia. Principes 29 (4). pp 166-169.

MOORE Jr. H.E. and UHL N.W. (1984). The Indigenous Palms of New Caledonia. Allertonia Vol. 3(5): 313-402.

MORAT Ph. (1986). Le Palmier que l'on Croyait Perdu. In L'Univers du Vivant, numero 7, page 44.

MORAT Ph.; JAFFRE T.; VEILLON J.-M. et MACKEE H.S. (1981). Les Formations Vegetales. Pl.15. Atlas de la Nouvelle Caledonia. ORSTOM Paris.

MORAT Ph.; VEILLON J.-M. and MACKEE H.S. (1986). Floristic Relationships of New Caledonian Rainforest. Phanerogams in Biogeography of the Tropical Pacific. Edit. Radovsky, Raven & Sohmer. Association of Systemics Collections and Bernice P.P. Bishop Museum. Honolulu, Sp. Public, numero 72: 71-128.

SCHMID M. 1973-1974. Les Palmiers de Nouvelle-Caledonie. Nature Caledonienne, numero 3: 22-25; numero 4: 14-18; numero 5: 409

Figure 82: *Basselinia deplanchei* in habitat (O).

FIJI'S NATIVE PALMS AND THEIR POTENTIAL FOR LANDSCAPING

R.H. (Dick) Phillips
Suva, Fiji

Lying as it does at the end of the Indo-Malesian chain, Fiji has a relatively small number of native palms, the majority of which are endemic. For the purpose of these notes I am including the Coconut *(Cocos nucifera)* and the Masei *(Pritchardia pacifica).* The former is so common throughout the tropical world that its importance is often forgotten - it is basically the Tree of Life to the Pacific and the low islands in particular could not have supported man had not the Coconut preceded him into the Pacific. The latter may have originated in Tonga but was a very early introduction to Fiji and is much more widely known than *Pritchardia thurstonii* which is limited to the islands of Lau to the east of Viti Levu. Indeed, it is only in the last few years that *P. thurstonii* has been available in any nursery in Suva although it is quite common overseas and I first saw it in the Ala Moana Park in Hawaii. There are no mature trees in Suva and seed supplies are intermittent.

One other palm, the Niusawa *Veitchia joannis,* is well known overseas but is still quite rare in gardens in Fiji.

The reasons for the complete neglect of native palms as suitable additions to the urban landscape are fairly simple. With few exceptions, native palms are widely scattered, often requiring long trips to get seed. In addition, they seldom make large colonies. The early administrators brought with them the plants and palms which they had known overseas so the landscape is dominated by *Roystonea regia, Livistona chinensis,* several *Ptychosperma* which have hybridised, *Chrysalidocarpus lutescens, Areca catechu, Phoenix dactylifera* and *Cocos nucifera.* It is only since the formation of commercial nurseries which started to appear about 1970 that palms have become more readily available for the ordinary gardener.

In addition to the palms mentioned above, a few individual gardeners have brought to Fiji an odd plant or two of palms which they had seen when overseas. This would account for about twenty palms which were hidden in private gardens. A few others have been imported for commercial testing.

In 1960 a ban on the importation of palm plants or seeds was invoked to prevent the introduction of any diseases which might affect the Coconut which is a vital item in the economic and cultural life of the Fijian people. This did not lead to a search for the local palms but only to a proliferation of the few common and readily available palms.

There are several references in Parham's Plants of the Fiji Islands that certain palms are in common use, but this is not so. For example, Parham states, ... "*Balaka seemannii.* A slender palm that is common in dense forests .... Widely used as an ornamental in Fiji and overseas."

B. seemannii occurs only on the islands of Taveuni and Vanua Levu - I doubt that I have seen more than twenty specimens on Taveuni and I would think that I am the only person who has this palm in cultivation in Fiji.

There are a number of different climatic zones in Fiji - the south-eastern sides of the larger islands lie across the path of the prevailing trade-winds and are wet and heavily wooded. Suva has an average rainfall of about 120 inches (3,000mm) per year,80 (2,000) in the summer and 40 (1,000) in the winter . The north westerly parts of the larger islands are drier and had lost most of their forest cover before the advent of Europeans. Nadi, where the International Airport is situated, has a rainfall of about 80 (2,000mm) inches per year, 60 (1,500) in the summer and 20 (500) in the winter . Most native palms grow in the wet zones.

Figure 83: *Neoveitchia storckii* in habitat, south-eastern Viti Levu, Fiji.

The majority of the native palms grow on the two larger islands, Viti Levu and Vanua Levu and most of the rest grow on the intermediate islands of Taveuni, Kadavu, Gau and Ovalau. Only the ubiquitous *Cocos nucifera* and *Pritchardia thurstonii* occur naturally on the smaller islands.

Several other points restrict the use of native palms in landscaping. The majority of the native palms occur in rain forests or in relatively heavily shaded areas; some occur at high altitudes (by Fiji standards) and will not grow and flourish at sea level. Many of these palms will not grow in open gardens, even in the wetter areas. In addition, the majority of the tourist hotels are situated close to beaches in the drier western areas and on islands off the coast from Nadi. In the latter cases in particular, the alkaline soils (pH about 8), the salt laden air and low rainfall makes the growing of native palms almost impossible.

Some relief from the high pH can be obtained by the use of mill mud, a product of the sugar process. Mill mud is the organic matter which is strained out of the sugar juice before it is clarified. It is about 95% humus and has a pH of about 5. It must be allowed to mature over a period of about six months as a large pile can generate great heat. I have used this mature mill mud with considerable success on the alkaline islands in hotel landscaping.

Cocos nucifera will grow under practically any conditions. However, it does best close to the sea and under 100 metres above sea level - it will grow at higher elevations and well inland but fruiting is usually poor.

Pritchardia pacifica is excellent in the wet areas and will grow very luxuriantly when given good soil, plenty of water, good drainage and just a little shade. It grows more slowly but very well on the dry alkaline islands with or without the addition of mill mud in the soil. To date, I have not tried *Pritchardia thurstonii* outside Suva but, considering the very hard conditions under which it grows on some of the limestone islands of the Lau Group, I would expect that it would do well on the alkaline islands near Nadi. I have not yet seen either species used on the dry side of Viti Levu - where some of the larger resort hotels are situated - but I feel that it would need considerable care and attention in the heavy soils during the dry season (the winter months).

Veitchia joannis is a palm which has considerable potential in the wetter areas but I have had trouble convincing gardeners and hotels that they should use it. The main criticism is that when the palm matures, the head of leaves is small in relation to the height of the palm - this criticism seems strange as it would take 30 or more years for the palm to reach such a height.

Metroxylon vitiense is excellent in very wet and swampy areas but it would be very difficult to get good results in dry areas even with heavy watering. There is also the added disadvantage that this palm needs to be replanted at about 15 year intervals.

Neoveitchia storckii is, perhaps, the palm that offers the most potential for landscaping and street planting. It grows in a very small area about 30 miles from the sea in wet lowlands. It appears to have adapted to the loss of the forest cover that young plants would have had. The heavy trunk is usually straight and the head of leaves is well proportioned even in very tall trees. Older palms often have whitish trunks which stand out well. In Florida, *Neoveitchia storckii* has proved susceptible to Lethal Yellowing. I have not yet seen this palm damaged by the Rhinoceros Beetle. I have sold and given away locally some hundreds of seedlings of this palm, but it is yet to become a regular feature of the landscape. I have tried to grow it on the drier side of the island without much success. Given a lot of water during the dry months and protection from salt winds it could be grown but few gardeners are willing to make the effort.

Most of the remaining native palms of Fiji would need shaded wet conditions in which to grow. Although some can be tall, most have slender trunks, some very slim, so for general landscaping they may not offer much scope. However, some may be very useful as indoor plants during the seedling stage and up to four or five years of age.

In particular, *Balaka* are very attractive as young plants and they are quite happy growing in small pots. However, they need shade and very humid conditions. As I now have no contracts for the maintenance of indoor plants, I have not been able to try these palms under air conditioning and believe that an effort would have to be made to keep the air around the plants as humid as possible. Some of the *Veitchia*

Figure 84: Crown of *Pritchardia pacifica* cultivated in the garden of R.H. Phillips, Suva, Fiji.

could also be of value as indoor plants and, as they tend to grow in stronger light than the *Balaka,* they may be easier to handle under normal indoor conditions.

Our two newest palms – *Alsmithia longipes* and *Gulubia microcarpa* – may offer some potential for landscaping but it will be some years before we can comment on this. The main population of *A. longipes* is at 1,000 feet in an area of very heavy rainfall – possibly in excess of 200 inches (5,000 mm) per year. However, there are several palms growing at sea level – seeds probably washed down a stream and growing along the bank. The rainfall in this area is high but the health of these trees indicates that this palm may accept conditions at sea level. These palms are those in my photos in Genera Palmarum.

Gulubia microcarpa is growing in an area where the forest has been logged at about 800 feet (280m) elevation. To date I have been unable to get seed to germinate and several seedlings taken from the bush lingered for several years before dying. Despite the fact that the seed is very small - about the size of a grain of rice - the palm is a large one and I feel it might grow at sea level so long as small plants were given shade.

Figure 85: *Cocos nucifera* on the east coast of Taveuni, Fiji.

VANUATU PALMS: THEIR DISTRIBUTION AND USES

Pierre Cabalion
ORSTOM, Paris
Translated by Marie Trefeu, edited by J.L. Dowe

Introduction: Vanuatu Plant Collections in the World

Knowledge of the Vanuatu (formerly New Hebrides) flora has developed in several stages, the last two Compendia published by Guillaumin in 1927 (1) and 1948 (2).

Since then many collections have been made; by botanists of the 1971 Royal Society Expedition (3); by ORSTOM (4), ethnobotanists and botanists from Noumea to 1979 and then from Port Vila; and in the field of Bryophytes by American scientists (5) in 1984 and 1985. No general and systematic work on the flora has been published since 1948.

In 1979, when I began my work on Vanuatu traditional pharmocology, I required good determinations for my samples. As information was to be found in numerous books and articles which often were not available in Vanuatu, this task was not easy. With a certain amount of innocence, I began to collect every published line on the flora of Vanuatu and compiled the literature under names of collector and identification of the specimens. At the same time, the Port Vila Herbarium (PVNH, proposed to be PVV) was developed to a normal standard sending three parts on four to other herbaria; one to Paris (P), one to Noumea (NOU) and one to Kew (K) or a specialist.

Although these two projects are not finished, we now have a better idea of the location of Vanuatu botanical collections in the world.

In the Pacific region, the richest institutions for Vanuatu specimens are Port Vila, Noumea, Sydney (NSW and SYD), Melbourne (MEL) and the Bishop Museum (BISH) in Hawaii.

The 1983 estimate for Noumea and Port Vila (6) Herbaria was respectively of six and 4000 numbers of Pteridophytes and Phanerogams. It is now of seven and 6000, amounting to a total of 8000 different numbers. Bishop Museum maintains some unique Vanuatu collections, for example Krauss's and Searle's specimens, and samples are classified by their geographic origin, which makes the work easy. Australia probably has the greatest number of exsiccata from Vanuatu, most of which have not been studied, as Australian botanists have had enough to do in Australia and Papua New Guinea.

I was unable to visit Melbourne; the herbarium there was curated before the turn of the century by Ferdinand von Mueller, who asked many missionaries and travellers to collect for him in the 'South Seas' (7). In Sydney (NSW), the estimated number of Vanuatu specimens is a total of two to 3000 Pteridophytes and Phanerogams. Fifty percent are represented by the Royal Society collections. Most of the other samples have yet to be studied: these were collected between 1880 and 1915, for example Quaife's and Layard's specimens.

Certainly, other herbaria in the world keep unknown Vanuatu collections. For example, in the United States, there are samples taken from Vanuatu by US soldiers during the Coral Seas War, when Vanuatu was base for operations.

A prudent estimate gives for Vanuatu the following results: the total number of collectors is more than 130. The number of Pteridophytes and Phanerogams exsiccata located in various institutions may be over 10,000 and 14,000 for Bryophytes.

I arrive now at my conclusion, which may lead to the preparation of the first Flora of Vanuatu. Technically the realization is now possible as there are about 10,000 samples for about 1500 indigenous and introduced species of Pteridophytes and Phanerogams in the archipelago. For Bryophytes, the figure is 14,000 specimens for about 600 species (8). Scientifically the conditions are now better than before, even if collections have not been made everywhere in Vanuatu. (For Pteridophytes and Phanerogams there is 0.7 sample/km^2, for Bryophytes there is one sample/km^2, algae has ten samples collected and lichens 200 samples). Botanists may now also find useful two new Floras, published or being published, for New Caledonia and the Fiji Islands, Vanuatu's closest neighbours.

VANUATU PALM GENERA

Genera thus far reported as occurring in Vanuatu may be grouped as such:

(i) Substantiated or recently collected
- *Licuala*
- *Metroxylon*
- *Calamus*
- *Gulubia*
- *Veitchia*
- *Carpoxylon*
- *Clinostigma*
- *Cyphosperma*
- *Physokentia*
- *Cocos*

(ii) Known or suspected to be introduced
- *Caryota*
- *Areca*
- *Pelagodoxa*
- *Roystonea*

(iii) Recorded but unsubstantiated
- *Livistona*
- *Chambeyronia*
- *Cyphophoenix*
- *Basselinia*

LICUALA

Licuala grandis H. Wendl. ex Linden

This species was once thought to have been introduced from New Britain and A.C. Smith reports that it occurs on Espiritu Santo and Malekula (A.C. Smith, 1979). Accordingly, Schmid reported it as occurring in the Rentabao Valley, very close to Port Vila on the island of Efate (Schmid, note added to the manuscript of 'Plant Species Observed on Efate', 1965). This is not indicated on herbarium sheets but, Mr. Robert Lambert, horticulturist of Port Vila, reconfirmed this to me in 1984.

The species also is reported on Pentecote (Schmid, Flora of Pentecote, 1974). A sterile sample, collected by Vienne indicates that the species is distributed in the Banks Group, where it has also been observed on Maewo by Genevieve Bourdy in 1986.

Non-Medicinal Utilization:

The leaves of *Licuala grandis* are used as traditional umbrellas or improvised in case of sudden rainfall, on Motlav (Dominique Bourret, No. 20NH), on Pentecote (Annie Walters, pers. comm. 1989) and on Malekula at South-West Bay (Pierre Cabalion, 16th November, 1982).

On Motlav, the leaf is used as a plate, as well as for wrapping traditional objects such as statues and masks which are created by inhabitants of Toman Island off the south of Malekula, as I observed at Port Vila during the Arts Festival of 1979.

This species is well known as a cultivated ornamental in many places throughout the world as noted by A.C. Smith (1979).

Medicinal, Ritualistic and Symbolic Utilization:

A medicinal use is recorded by G. Bourdy in her thesis. '. . . On Maewo . . . the leaves are used to cure the condition known as GWATUWAWALE. This is a 'headache' which affects men who have been in contact with menstrual blood. The leaf must be crushed as it is very tough and the pulp thus obtained is mixed with a little water, so as to make a juice which is spread on the head of the patient. The treatment is repeated daily until recovery . . .' (G. Bourdy, 1989, Annexes.) This of course is only symbolic.

In the central region of Pentecote, which is an APMA language area (Tryon, 31, 1976), the leaves of this species represent a 'grade' within the social structure assigned to women. In fact the society of northern Vanuatu is structured so that everyone is required to pass through grades to attain higher status. The *Licuala grandis* leaf is specifically symbolic for a woman who has passed the initial grade or RABANKAMEL. After attaining the pass into the next grade, the woman will always have the leaf, either in her hand or slipped into her belt, at the time of passage into subsequent grades, so as to indicate that she is a woman who has successfully reached that status. (Annie Walter and Pierre Cabalion, 2nd October, 1984).

Vernacular Names: (as per the enumeration of languages by D.T. Tryon, 1976).

– NDAUBIHU on Maewo, from the TETERARA language (T, 20), (Thesis of G. Bourdy, 1989).
– NEYEDIP, a language from Mota Lava (T. 5), (D. Bourret, NH 20).
OVLOR TATA, OVLOR is the name of the species on Mota (T, 6) and TATA denotes the natural occurrence (non cultivated) of the species (D. Bourret and B. Vienne, pers comm).
– WILOG WEWIA is a name mentioned in herbarium notes by B. Vienne for a sample collected in 1972 in the Banks Group (T, 3 to 18)
– TABATABA in the north of Pentecote, in RAGA language (T, 30), according to M. Taillade, ex-Professor of Art at Bougainville School, at Port Vila as well as APMA language (T, 31) (Annie Walter and Pierre Cabalion, 2nd October, 1989) To be more precise, BUTSU TABATABA indicates the arborescent character, BUTSU meaning 'the trunk.'
– SONGE M JAM in the central south west of Pentecote, according to Schmid (Flora of Pentecote, 1974) and probably in SEKE language (T, 32).
– PUNE PELE, in PIAMATSINA language from the Cumberland Peninsula of Espiritu Santo.
– NIRIVIUH in the BENOUR and LEMBINWEN languages from South-West Bay of Malekula (P. Cabalion, 16th November, 1982).
– NEVEHU at Lamap, south-eastern Malekula, according to M. Taillade, probably in the Port Sandwich language (T, 81).

METROXYLON

Metroxylon warburgii (Heim) Beccari

Metroxylon salomonense (Warb.) Beccari

Metroxylon is reported on Anatom as being "... near the houses..." by MacGillivray in 1853 (1854) and by Guilfoyle in 1968, in which he states that this palm was " . . . abundant in the interior of the island . . . " (Guilfoyle, 1869). This is not the case today, as the only villages are near the seashore and the interior is completely deserted.

Guillaumin reported it in 1937 and 1948 under the name of *Coelocarpus* and considered it endemic to Espiritu Santo and Malekula. Schmid reported it on Efate in 1973 (Vegetaux Superior of Efate), on Tanna (Flora of Tanna, Supplement I, 1970) where it is occasionally planted and on Erromango (1974) where it is also cultivated. Although the only species thought to occur in Vanuatu was *M. warburgii,* it has recently been reported by John Dowe that *M. salomonense* is also found (see page 30 of this volume) in the archipelago.

Notes:

This is a monocarpic genus, and therefore the palm dies following the first fructification. The inflorescence, positioned vertically at the apex of the trunk, becomes yellow when the fruit is ripe. The leaves also turn yellow and finally the entire palm dies. As with the agaves or the *Cerberiopsis,* the decline and decay is very long, exceeding six months, as I have witnessed at Port Vila.

The fruit is decorative and appealing and germinates very easily. Nevertheless, this palm is not regarded as an ornamental and is avoided due to its spiny petiole and rachis.

It would be of interest to determine if the native peoples of Vanuatu differentiate between the two species, if they are named vernacularly or if they have different uses. Further study is needed in this regard.

Utililization:

There appears to be little use of *Metroxylon* as a medicinal plant in Vanuatu. Only one reference is known: Annie Walter reports (AW 225 and 225A) that on Pentecote, the juice of the stem is used in the treatment of diarrhoetic bleeding. In fact this is medically inconsistent, as it is often the result of the presence of intestinal parasites, such as 'amibes' or giardia, and these parasites, at least during in vitro cultivation, require the presence of starch similar to that found in the *Metroxylon* juice.

The use of *Metroxylon* as a food source is well documented, for example in Papua New Guinea. According to the available literature, the use of *Metroxylon* starch is only a recently introduced practise in Vanuatu (Guillaumin, 1954). Claude Jardin mentions that *Metroxylon warburgii* is used as a food plant in Vanuatu, but without further details. Schmid, in his Flora of Erromango, says that the species is cultivated as well as occurring naturally. According to Jean Guiart, the starch is extracted in remote localities. Barrau notes that on Pentecote and on Tanna, the extraction of starch can be encountered (Barrau, 1956). Such brief mentions encouraged me to seek more information on the use of the starch of this palm in Vanuatu. On Pentecote, no records of the edible uses have been recorded (P. Cabalion, 3rd March, 1979. Annie Walter, pers. comm.)

During a visit to South-West Bay, Malekula (16th November 1982) I was informed that *Metroxylon* was not a food for human consumption, " . . . befo man ino kakai insaet blong hem . . . ", but that the base of the trunk was used for pig food. According to an inhabitant of Fila Village, he saw, on Espiritu Santo, *Metroxylon* being used as a source of starch as well as a source of salt. Some 'Bushmen' from the interior of the island had cut down a large palm and removed the inside of the trunk. The removed material was crushed, placed back in the hollow trunk which was then filled with water. After some time the removed material sank to the bottom, and the 'water' was siphoned off into a 200L container where it was left to ferment. Edible starch was obtained after the removal of floating salty matter which was later used as a condiment in cooking.

Starch has been obtained from *Metroxylon,* primarily by the population which lives in the interiors of the islands, for a long time. At times, it has also been a source of salt due to the difficulty of travel to the sea, usually brought about by 'wars' between the 'MANBUSH' and those people who live by the sea, known as 'MANSALWATA.'

It appears that the gathering of starch from *Metroxylon* is declining, with the gradual introduction of other starch rich foods (for example manioc) which are easier to cultivate. Even other 'traditional' starch sources, such as yams and taro, are becoming less cultivated. According to Barrau (1956), the presence of *Metroxylon* in Vanuatu is due to its introduction by man as a food source. This raises the problem of evolution of agriculture in Vanuatu. On Anatom, the population is presently found almost exclusively living near the coast, and, contrary to the observations of Guilfoyle in 1868, which state that the island had many thousands of inhabitants scattered in small villages in the mountainous interior, today there are very few *Metroxylon* to be found in the interior. This may suggest that it has been introduced to Vanuatu and is not the natural environment.

On Pentecote, an unusual use has been observed; the base of the petiole is used to grate bananas and yams in the preparation of LAP LAP, a sort of pudding cooked between two stones.

In the artistic domain, the hollow trunks are used as large dye tubs in the preparation of ceremonial matting.

The most important contemporary use of *Metroxylon* is for construction material. The leaves are used for the fabrication of very durable thatch used as roofing in traditional houses. In central Pentecote, the leaves are used for making tiles which are supported on beams. Similar tiles have also been reported from Vanua Lava (P. Cabalion, 1983). In northern Pentecote, it has been observed that the rachis, cleaned of the leaflets, is used as a rafter (Loltong, P. Cabalion, 8th March, 1983).

On Malekula, at South-West Bay (P. Cabalion, 16th November, 1982), at the villages of Mele and Fila near Port Vila (P. Cabalion, 18th January, 1982) and on Erromango, it has been observed that the leaves are used as a durable thatch, reputed to last fifty to sixty years, and superior to the thatch made from WAELKEN (an unidentified Poaceae). *Metroxylon* is specifically cultivated for this purpose (P. Cabalion, 21st June, 1982).

Although not usually mentioned in notes, the leaves are used in the making of baskets and matting. This particular use is very common on Pentecote (P. Cabalion, 3rd May, 1973).

Vernacular Names:

Metroxylon has many names throughout the archipelago. On Erromango, there are two names in use, which may indicate that there may be two species present. The names are ULUWAR and NEVREPIN which suggests that the plants are identical in trunk but different in leaves and fruit (P.

Cabalion, 10th November, 1985). NEVREPIN and NUVREIPIN are used in the south of the island and elsewhere the equivalent NEMPREHI, which indicates the same species in the traditional sense (P. Cabalion, 26th June, 1982). On Pentecote, the only indigenous language still used is that of the SEE language, although it appears that many plants retain their names from the extinct languages of URA and UTAHA. In 1985, there were only five people still alive who could recall the URA language. To demonstrate the abandonment of vernacular names in contemporary Vanuatu, a questionaire was submitted to students in grade 8 at the Bougainville High School. Two lists of words were given to the students, the first of vernacular words the second of contemporary words which were their equivalent. Few students could match the names from the second list to the first list, which of course indicates a lack of knowledge of vernacular words.

In conclusion, it appears that *Metroxylon* is distributed throughout the entire archipelago and has been that way for some time as most children know at least one vernacular name. Unfortunately this is not the case with other plants in Vanuatu; the children are 'forgetting' more and more the names through the structure of the contemporary education system which separates children from their villages for long periods.

It is by no means that the ethnobotanical study of *Metroxylon* is complete, although it is nevertheless legitimate to speculate that it has been known and utilized by the inhabitants for a long time. In my opinion, *Metroxylon* has been introduced to the archipelago by man, firstly as a source of food (although not used for this since the 1960-70s) and secondly as a source of building material, primarily as thatch for the roofing of houses. It is probably an example of one of the most ancient introductions of plants into Vanuatu, equivalent to the contemporary introduction of taro and yams. Analysis of myths and legends and archeological exploration may prove this.

CALAMUS

No botanical revision of this genus has been undertaken in Vanuatu, although there are reports of species being distributed throughout the archipelago.

On Efate, D. Levat in 1883 reported LELAOROS (T 157) as occurring inland of the island (Guillaumin 1919). This species appears to be identical to the one found on Espiritu Santo by Schmid (1965). There is the same indication by Schmid (1973, Veg. Sup. Efate: 13) which indicates the species is abundant in the forests of the interior plateau. Also in the south/centre of the island at the edge of the secondary forest (Schmid Phaner. NH: 51), Schmid 3271, NOU:RSNH 1076, PVNH; S. Seoule 38, 12th August 1982, PVNH, NOU and P; R. Clarck 13, by Ms G. Bourdy, 9th May, 1987, PVNH.

On Erromango, *Calamus* sp. has been collected many times by Schmid. On 12th May, 1970, 3272, NOU; by the Royal Society Expedition, 3rd October, 1971, 118, PVNH; by P. Cabalion, 3038, 16th November 1985, PVNH, NOU and P. On this island, *Calamus* sp. is found in the forest at about 250-300m. (Schmid, 1973, Phaner. NH: 51), common in the forest on basalt in the south-east and centre at about 200-300m (Schmid 1974, Erromango: 50)

Calamus is not reported from Tanna (Kajewski 1930).

On Espiritu Santo, in forest on the plains surrounding Big Bay, where it is sparse and not forming any obstacle in the undergrowth (Sarlin 1950). It is not harvested.

On Pentecote, *Calamus* sp. is noted by Schmid (1974): as an armed climbing palm, common in forest at 300m; *Calamus (rotang)* was collected by Phillippe Morat, 25th April, 1977, 5470, NOU; a *Calamus* sp. by Pierre Cabalion, 3rd October, 1980, 1153, known as ROTIN, KAULA and KAIKAULA (T 31), NOU, is similar to Schmid specimens 3271 and 3272.

On Motalava, Mrs. D. Bourret collected at 300m a *Calamus* sp., 216, NOWOL (T 5), May 1978, NOU: in the Banks Group, B. Vienne collected in 1972 a *Calamus* sp. NOU.

Utilization:

Medical Use: Ointment

At Loltong, in the north of Pentecote, the sap from the stem is used as an ointment to calm ear irritations when there is no presence of pus (P. Cabalion 1153). This resembles the utilization of physiological serum in modern medicine, with no apparent healing objective.

Non-Medical Use: Fibres, Sap, Timber

Still at Loltong, the spines, present under the leaves of *Calamus*, are used for fishing hooks to catch flying fish. They are attached to the fishing line with a yellow thread made from spider-web, commonly found throughout Vanuatu near habitations. The spider is known as MANGARU at Loltong (T 30). The local name reflects the palms' utilization: GARU-FUALAI (T 30), means hook plant, ROTIN in local French and LAIKEN or LAWYER CANE are used in bichlamar and local English.

The stems are used for the construction of houses and huts. The sap of *Calamus* is appreciated whilst in the forest to quench the thirst. One only needs to sever a stem and drink, similar to the stem of *Entada phaseoloides* (Mimosaceae). In the interior of Pentecote, the fibres stripped from the stems are used for the fabrication of cord used to secure the handles of tools and facilitate their control (P. Cabalion and A. Walter, 28th Sept. 1984, Enkul Village).

On Erromango, this palm is not consumed, but used in the preparation of cord for construction (P. Cabalion, 10th Nov.1985).

The fibres of variety TRU NORMA (the small one) are used because they do not break readily whereas the variety TRU MORENI (the large one) breaks easily (P. Cabalion and G. Bourdy, 3028, Happy Land, 10th Nov. 1985 and Ipota, 10th Nov. 1985). The vernacular name TRU (T. 161) covers two botanical species. *Calamus* sp. is considered the male whilst *Flagellaria indica* (Flagellariaceae)(P. Cabalion 1719) is considered the female. Both are climbing plants which possess leaves and organs which become attached to anything that comes across their path, either by hooks (as with *Calamus)* or tendrils (as with *Flagellaria*). The name *Calamus* in local bichlamer is LOYAKEN (P. Cabalion, 19th June, 1982).

Another use for *Calamus* stems has developed recently in Vanuatu, and now there are workshops where rattan-cane furniture is constructed and sold locally. Armchairs, beds, chairs and tables are made in small quantities and at present does not endanger the natural population of *Calamus.* Possibly, if the exportation of these articles becomes more important, it may be necessary for the Service des Forets de Vanuatu to take action before it is too late.

In conclusion, the genus *Calamus* in Vanuatu is in need of study by palm botanists, although they usually prefer to work with less prickly problems and unarmed genera. The traditional utilization does not endanger the genus; we will be able to continue to drink from time to time the sap of this palm in the forest, or use it as an ointment. The people of the villages will continue to use the fibres to make cord and the spines to make fishing hooks. The only potential danger could come from the excessive collection of stems for the manufacture of rattan-cane furniture.

GULUBIA

Gulubia cylindrocarpa Beccari

This species has been collected on Vanua Lava, Malekula, Erromango and it can be supposed to be well represented on most of the islands of the archipelago.

On Malekula, at south-West Bay, the fruit, about 2cm long and 0.7 - 0.8cm in diameter, ripen in mid-November. At this particular location, the infructescence appeared to bear only limited fruit (Cabalion 1819, 15th November, 1982).

Utilization:

The 'heart' of *Gulubia cylindrocarpa* is comsumed and the fruit is eaten by children. (Malekula, specific mention).

Vernacular Names:

At South-West Bay, in the Benour and Lembinwen languages (T 115 and 116 (1976)), this species is known as NIULIP, the derivation 'NIU' meaning palm (Malekula, specific mention).

VEITCHIA

Veitchia arecina Beccari is known from Tanna (Schmid 363) and Nguna (P. Cabalion 1360). In his Phanerogams of Vanuatu (Schmid 1973) he mentions its location on Tanna as being south of the volcano. *Veitchia joannis* Wendland is mentioned by Guillaumin in his Compendium (1948). His reference is to Seemann who lists it in his Flora Vitiensis (1873).

Veitchia macdanielsii Moore was collected on Espiritu Santo by MacDaniels (9108 and probably 3108) on 29th December 1949. The type is at BH. The formal description was completed by H.E. Moore Jr. (1957).

Veitchia metiti Beccari is a species from Vanua Lava. Its traditional uses are not known. It is mentioned by Beccari (1925), Guillaumin (1927) and H.E. Moore Jr. (1957).

Veitchia montgomeryana Moore is another species described by Moore (1957).

Veitchia spiralis Wendland was first collected on Anatom by MacGillivray (Seemann, 1873: 2670) and then by Kajewski (Guillaumin, 1932) and described under the synonymous name *Kajewskia aneityensis.* T.C. Whitmore lists a specimen on Efate (fruit at PVNH and K) which extends the distribution of the species toward the centre of the archepelago. This species is mentioned in passing in articles by Guillaumin (1948), Burret (1932), H.E. Moore Jr. (1957) and Schmid (Flora of Anatom, 1970). A.C. Smith made an implicit allusion to the specimen of Kajewski (1979).

Veitchia winin Moore was another species described by Moore. There are at least five specimens; on Malekula (MacDaniels 3093 and 3076, and RSNH 306), on Erromango (RSNH 16263) and possibly on Pentecote (P. Cabalion 2592).

Notes:

It is of interest to add to the above species some undetermined specimens mentioned by Schmid in his diverse Floras and other authors. On Efate . . . 'an Arecaceae, *Veitchia,* with tall, slim trunk up to 20m tall, appears to be distributed throughout the entire forested areas. It is most common on the crests and visible from far away, the palms taller than the surrounding forest . . .' (Schmid, Vegetaux Superior, Efate, 1973).

(An attempt at germination of seeds collected from a specimen from 550m alt on Efate (P. Cabalion, 1274) resulted in failure. Numerous *in situ* germinations of a *Veitchia* sp. have been observed on Nguna, on the slopes of the volcano in protected areas where humidity is constant and moisture maintained by temporary creeks. Less evidence of germination was noticed near the summit at 500 to 593m alt. It is apparent that this species is left standing during land clearance on the island. (P. Cabalion, 1360.)

Another *Veitchia* sp. with less elongated fruit grows at low altitude according to Schmid (Vegetaux Superior, Efate, 1973).

On Tanna, the indigenous people distinguish two species of *Veitchia,* known as NA FUA and NA PUAK. NA FUA has 7-9 leaves about 2m long, with the petiole naked for about 40cm.; annular rings are not prominant, much elongated in shady areas. Both species occur in the forests in the interior of the island, although NA FUA is found in the Lenakel area on coral platforms and around Port Resolution. The distinguishing characters of the two species have yet to be studied (Schmid, Flora of Tanna, Supplement I, 1970).

In Phanerogams of Vanuatu (Schmid, 1973), a *Veitchia* (cf. *V. arecina?*) is mentioned as occurring on Tanna . . . ' on slopes south of the volcano' . . . It is surely NA FUA mentioned in 1970, as located on the slopes of the volcano.

On Anatom '. . . a tall palm, only encountered in a small population in upper Umec, south-east of the island . . .' (Schmid, Flora of Anatom, Supplement II, 1971).

Utilization:

According to Sheila Gowers (1976), the endosperm has a consistency of gel and is edible. This has not been confirmed for *V. arecina* on Nguna (P. Cabalion, 28th September, 1979) nor for the three specimens collected on Pentecote (Annie Walter 80 and P. Cabalion 2592). The apical bud of the latter sample is edible, having an agreeable flavour similar to ripe, fresh hazel nuts. It is regrettable that such magnificent palms are cut down for this purpose (Pentecote, 28th September, 1984).

On Maewo, according to Genevieve Bourdy (708), there is a *Veitchia* sp. collected at 350m alt near some gardens, which has an edible fruit consumed when immature and still green; this species changes from yellow to red at maturity.

On Vanua Lava, the trunk of a *Veitchia* sp. is used for planks. This is the specimen P. Morat 7447. It is certainly indigenous to the island, and not introduced, as it occurs only in primary forest (P. Cabalion, Vanua Lava, 14th June, 1983).

On Erromango, there is a *Veitchia* sp. referred to as WAEL COCONUS (wild coconut) because of the basic resemblance to that palm. The apical bud is consumed, the trunk split for use in the construction of partition walls in huts and the leaves used in the fabrication of baskets. It is still very common on Erromango, being a magnificent tall palm (P. Cabalion, 2264, 25th July, 1983).

It appears that the leaves of *Veitchia* are as much appreciated as those of the coconut palm for the fabrication of baskets as well as being considered superior to those of *Clinostigma harlandii* (P. Cabalion, Efate, 8th June, 1985).

Vernacular Names:

WOGHA MI IR – On Vanua Lava, the dialect from near Sasar language No. 8, P. Cabalion, 14th June, 1983, from P. Morat (7447).
TOTOU GAT UTU – literally 'tree of the pigeons', on Maewo, Saritamata, either languages Nos. 20, 21, or 22, Genevieve Bourdy 708.
NAPUPUA or NAPUA – On Nguna, language No. 152, P. Cabalion, 28th September, 1979 and 31st December, 1981.
NIVFAL – On Erromango, Sie, language No. 161, P. Cabalion, 2264.
NIVAL – On Erromango, Sie, language No. 161, P. Cabalion. 10th November, 1985, Chanel Sam 247 and Genevieve Bourdy 218.

CARPOXYLON

Carpoxylon was listed by Schmid as occurring in the forests in the interior of Anatom (Flora of Anatom, 1970). That author expressed doubt about the distribution of the genus by marking it with a '?', and correspondingly there is no herbarium numbers allocated by Schmid to *Carpoxylon.*

It is with interest that I was recently notified by Mr. John Dowe of his rediscovery of this genus elsewhere. (See page 106 of this volume)

CLINOSTIGMA

Clinostigma harlandii Beccari

Two recent collections are known from Vanuatu.

– From Vanua Lava, 11th June 1983, Phillipe Morat 7419; above the sulphur mines on the left bank of the Chelva River at about 550m alt. parts preserved at PVNH, NOU and P; determination by John Dransfield, July 1983 as *Clinostigma* cf. *harlandii.*

– From Efate, 8th June 1985, Pierre Cabalion 2722, Mt. MacDonald at 600m alt., near the point marked 636m on the 1/50.000th map. The species appeared at 500m alt., below which it is replaced by a *Veitchia* sp. Inflorescences (four) crowded with fruit (P.C. 2712), spathe 1m to 2m; palm about 15m tall, 15cm in diameter, leaves 2m long, with petioles 50cm. Germination was attempted but without success. Conserved at PVNH, NOU, BISH, P and K. Quoted by John Dransfield in a letter from P.S. Green, 5th December 1985 to the author.

Taxonomic Notes:

Moore and Fosberg placed in synonymity all the species of *Exorrhiza* under *Clinostigma* in Gentes Herbarum 8: pp. 458-466, although, in 1956, Burret maintained the separation.

Another case of synonymity is to be noted for *Clinostigmopsis* of which one species had been described by Pichi-Sermolli for Vanuatu. Burret maintains this genus as *Exorrhiza* although it is synonymous with *Clinostigma.*

It appears that all is still not resolved as Willis, ed, (1985) mentions two species of *Clinostigmopsis,* one from Vanuatu and one from Fiji. They are sometimes included in the genus *Exorrhiza* (p. 262) although only one species of *Exorrhiza* exists (p. 45) – the one from Fiji. This confusion indicates the need for a revision of the genus.

Utilization:

The palm-heart and fruit are consumed from this species, although the quality is inferior to that of *Veitchia* spp. and the coconut palm.

Clinostigma spp.

Several undetermined specimens of *Clinostigma* are known from Vanuatu. Collections by researchers from Noumea indicate that this genus is well distributed throughout the archipelago.

On Anatom, Schmid noted the presence of *Clinostigma* sp. (Supplement to the Flora of Anatom, 1976). It is described as a tall palm resembling the coconut palm, but having leaves with a horizontal arrangement, numerous adventitious roots up to 60cm long and very small fruit; collected west of Ougapnaerek at about 400-500m alt., locally abundant but known from only one location, situated in a secondary range parallel to the main range on the island. The same observations were also included in Phanerogams of Vanuatu (Schmid, 1973) – *Clinostigma* sp. western Anatom, Ougapnaerek, in forest about 400-500m alt. The specimen Schmid 3794, 16th May 1971 would probably be for the above reference. At NOU, and inflorescence of *Clinostigma* sp. is displaced, and is possibly from Schmid 3794. Many other parts are also conserved at NOU.

A second specimen is Schmid 3795, 14th May 1971; Juvenile *Clinostigma* sp., west of Ougapnaerek. It is conserved at NOU.

On Erromango, in his Flora of Erromango, Schmid notes that there occurs what appears to be the same *Clinostigma* sp. as is found on Anatom and Pentecote (Schmid, 1974). He indicates that the *Clinostigma* sp. from Erromango is a tall palm with a robust trunk, the base surrounded by adventitious roots 1-2cm in diameter, inflorescences with many branches, thin and stiff, and with very small fruit; growing in forest on basalt, inland at about 450m alt. There are two collections by Schmid which correspond to the description.

- Schmid 4777, 26th June, 1973, central Erromango at about 400m, parts conserved at NOU.
- Schmid 4778, July 1971, central Erromango at about 500m, one part conserved at NOU.

One other collection also appears to have been used in his floras, although the generic determination is a little in doubt. It is:
– Schmid 3273, 12th May, 1970, south-east Erromango, *Physokentia* or *Clinostigma,* one part conserved at NOU. In his Flora of Pentecote, Schmid notes (p.2) the presence on the island of " . . . different palms with at least one species, *Physokentia* sp. found at about 500m and belonging to a genus which does not appear to be represented farther south . . " Erromango is farther south than Pentecote Island. Accordingly, Schmid 3273 could be *Clinostigma.*

On Pentecote Schmid notes ". . . at over 600m there is a stunted forest which is locally dominated by tall palms (*Clinostigma*) (Schmid, Flora of Pentecote, Jan, 1974). On the following page he states that ". . . *Clinostigma* is most frequent at relatively high altitude . . ." On page 24, Schmid describes *Clinostigma* sp. from the sample which he collected as: ". . . large palm with the trunk at the base surrounded with a thick cone of adventitious roots which act as stilts . The trunk is 30cm in diameter, leaves held horizontal, regularly pinnate. The inflorescences are composed of many branches, arranged in groups, with small fruit less than 1cm wide, a beige elongate spathe. It was observed at more than 500m alt., abundant on crests. It appears identical to the *Clinostigma* of Anatom and Erromango . . ."

The sample collected by Schmid on Pentecote is as follows: – Schmid 4696, 17th June, 1973, Pentecote 500-800m, one part with fruit at NOU (Locality inexact, but 'Bavaret sector', from his Flora, p.2) This locality is near the airplane runway, south of Melsisi.

On Espiritu Santo, an ethno-nutritionist collected a palm described as *Clinostigma* sp. for Mrs. Dominique Bourret and the author, for their research. It is:

- Igor de Garine, 21st Nov, 1977, Santo, Buria, conserved at Port Vila.

For Vanuatu in general, Van Steenis et al mention in 1966, the presence of the genus *Clinostigma* in Vanuatu (pp. 146-147), whilst A.C. Smith mentions it in 1979 (p. 429)

Utilization:

On Espiritu Santo, the fruits, as large as marbles and containing an 'almond' are consumed. It is known as OLO-OLO PATUB in the interior of the island (Igor de Garine, 40). On Pentecote, the *Clinostigma* sp. collected by Schmid (4696) is named KOLKOL (Schmid, Flora of Pentecote, 1974) This is most likely Seke language, listed as 32 by Tryon (1976).

CYPHOSPERMA

Cyphosperma cf. *balansae*

First collected on the Cumberland Peninsula, Espiritu Santo, this species has been determined by Dr. Natalie Uhl to be *Cyphosperma balansae,* although considered endemic to New Caledonia. Further collections and examination should be undertaken to verify the determination (see page 145, this volume).

If in fact this species proves to be *Cyphosperma balansae* one can imagine, with some reservations, that it may have been introduced into Vanuatu more or less recently. Many Ni-Vanuatu people have worked in New Caledonia and the species could have been noticed by one of the workers . . . only one seed would have been necessary.

PHYSOKENTIA

Physokentia tete (Beccari) Beccari

This species was first collected on Vanua Lava by Harland (sometimes spelt Harlandt) and mentioned by Guillaumin in 1927 and 1948 with no indication of traditional uses. Maurice Schmid has also collected it on Pentecote and mentions it is his Flora (1974). He also mentioned the presence of a *Physokentia* sp. in the south-east of Erromango (Phanerogams of Vanuatu, 1973). This is probably his sample 3273, conserved at NOU and labelled as *'Physokentia* or *Clinostigma.'*

The distribution of this genus will be clarified with further collections. For the time being, the presence of *Physokentia tete* on Vanua Lava and Pentecote is certain, and possibly on Erromango, and wide spread throughout the archipelago.

COCOS

Cocos nucifera L.

It appears that botanists did not bother to collect this species, presuming that the agronomists had volunteered. From my research, it appears that there is only one specimen preserved in Paris (P), it being: Lequerre (Admiral), sometime prior to 1927, at Port Vila, according to Guillaumin (1927).

Nevertheless, the presence of *Cocos nucifera* in Vanuatu is mentioned in several articles.

On Anatom, the coconut palm is one of four palms observed by MacGillivray in Nov. 1853 (see *Caryota)*

On Efate, Guilfoyle was surprised to find a coconut palm is a village in the interior of the island ". . . the coconut will seldom grow well far away from the sea, although on the island of 'Vate', one of the 'New Hebrides' I found several fine specimens in a village about nine miles from the coast . . ." (Guilfoyle, 1969). Perhaps it is the palms near Port Havannah mentioned for their beauty, ". . . groups of palms and tree-ferns of great beauty . . . gave to the landscape a peculiar charming effect, such as I had not previously witnessed in other islands, but in parts of 'Vate' . . ." (p. 131)

The locations where Guilfoyle found many villages are today entirely uninhabited, but the disturbed vegetation and sometimes rows of stones around a level surface, indicate previous inhabitation by man; for example around Mt. Erskine and on the hills above the Matoua and Teouma Rivers.

On Efate, Guilfoyle once again mentions coconut palms growing amongst the coastal vegetation of the Bay of Pango (p. 133).

On Protection and Deception Islands, also known as Moso and Lelepa Islands, in the Bay of Havannah, Guilfoyle observed several coconut palms used as a substitute for water.

" . . . during my rambles upon these islands, I did not meet with a drop of water. As a substitute for water, the natives use the milk of the coconut which may be had in abundance . . ." (p. 132).

The production of copra was the first industry developed in the archipelago (Guillaumin, 1954). It is of interest to note that cattle were first introduced into Vanuatu as a means of clearing the plantations of excessive undergrowth, but nowadays beef production far exceeds the value of copra production. The coconut plantations have passed their most productive years and the maintenance and introduction of high quality varieties of coconut palms have not eventuated (Coconut Research Station, 'IRHO',

Santo). It appears that the copra industry is in such a state of decline that one could say that King Copra has abdicated his throne, possibly never to receive it back again.

On Espiritu Santo, John Baker noted the presence of coconut palms up to 1400 feet alt. at a location north of Mt. Tabwemasana (Baker, 1935).

On the northern islands of Vanuatu, the 'wild' coconut palm is a source of copra which is sold to traders. "... quite a lot of wild coconuts are gathered and made into copra by the natives and this is picked up by cutters and small schooners ... " (Kajewski, 1930). The description as 'wild' may not be entirely true. In fact, in many local plantations, the palms are rarely planted in lines and therefore does not mean that their presence is due to coincidence.

On the southern islands of Vanuatu, the coconut palm is abundant, although on Tanna in the vicinity of Yasur volcano, great numbers were killed in early 1988 due to its eruption. On Erromango, the coconut palm is less abundant and grows only in numbers sufficient for local consumption and not for copra production. On this island, one of my colleagues, Peter Lomo, planted a coconut palm on Mt. Fedmoghum at 758m alt. in commemoration of our visit to the area. I wish to mention this to avoid later confusion in regards to the true distribution of the species.

The role of the coconut palm in Vanuatu and its exact distribution throughout the islands of the archipelago deserves a detailed study, which unfortunately, exceeds the plan of this present article. A thorough research was initiated in 1985 by Mr. Barry Weightman, ex-Chief of Agricultural Services, Port Vila. Previously, research findings have been published by Jacques Barrau in 1954, according to Guillaumin.

Traditional Utilization:

All parts of the coconut palm have a traditional use; for timber, fibres, diverse extracts and the consumption of its fruit and heart.

Many aspects of everyday life in the villages throughout Vanuatu depend more or less on the coconut palm. The trunk is used as is other species of trees; the leaves used to create diverse objects, usually of a 'throwaway nature', such as baskets for transporting fruit, vegetables and animals; for making roughly woven hats; for temporary fences during special occasions and fetes and even torches of dried leaves. The multitude of uses could fill an entire chapter.

On the island of Erromango, the chiefs announced their presence on the battle field by attaching, to their upper arms, hollowed coconuts into which are inserted erect stems of a grass, *Imperata cylindrica* (NAMKAI, P. Cabalion 2085, PVNH, NOU, and P.) This display allows the 'Big Men' to be more easily seen and noticed (Ipota, 21st July, 1983).

The floss of the coconut is used to spin threads which are then woven to produce strong though short cords. This in turn is used to make strong, elegant and functional combs – the small cords bind together the teeth which are the lignified parts of tree fern trunks.

Throughout Vanuatu, the custom of 'Pig Ownership' is very important. Coconut palms are an important ingredient in the preparation of pig food. For example, on Erromango, the leaves and bark of *Tylophora aneityensis* (Asclepiadaceae) are crushed together with grated coconut endosperm to be given to pigs for fattening (Ipota, 25th July, 1983).

As human food, the heart of a young coconut palm is preferred to that of *Clinostigma harlandii* (Efate, 8th June, 1985).

Rituals and Magic:

Parts of the coconut palm are often used in medicine or in magic philtres, administered according to the symptoms which the patient displays. On Pentecote and Ambae, the liquid from the coconut seed is the materialization of the action of the healer or the one who knows the necessary incantations. According to Codrington (1891) " ... on Pentecost and Leper's Island, the juice of a very young coconut, on which the doctor has blown, with a charm muttered or sung, is drunk by the patient or rubbed upon him ... "

Also on Pentecote, the smoke from burning dried coconut serves as a conductor for treatments. From Codrington (1891), " ... On Pentecost, if a man is delirious they say a mae, that snake of mysterious nature, is in his stomach. A doctor will then breathe his charm into a dry coconut husk which he has set on fire; the patient sits over the smoke, and the snake, which is a ghost or spirit, is driven out ... "

On Tanna, an example is given by Guiart (1956) " . . . at the time of planting the crop of ignams (yams) . . . they are sprinkled with the milk of young coconuts, then with seawater and then with water in which the juice of a certain (determination unknown?) plant has been squeezed . . . in the temple, temporarily set up in the middle of the field, plaited coconut leaves are laid down upon which are placed the shells of coconuts and over which the concoction is sprinkled . . ." Guiart (1956). This is a further example of the coconut palm being used as an indirect support or materialization of the magic action which will ensure the protection and success of future harvests.

On Erromango, an interesting example of the coconut palm used as a symbol of healing bone fractures is noted. Expressly for the purpose, a germinating coconut is planted in the ground, the first leaves being surrounded by a bundle of stems of a species of Poaceae, *Chrysopogon aciculatus* (Retz.) Trin. Apparently, when the palm commences to grow, the bone is healed (Peter Lowo and James Atnelo, Erromango, 25th July, 1983).

Medicine:

In these examples, the coconut palm enters the preparations as either an active or passive element.

On Tangoa, I obtained two recipes to prepare a remedy for 'pis-pis yellow.' This symptom corresponds to a colouration judged abnormally strong of the urine, the result of many causes, such as gonococcis, etc.

For the preparation, it is necessary to use a 'visible' equivalent, that is a variety of coconut palm with yellow fruits. In the first recipe, roots taken from such a palm are cut into pieces, softened by exposure to heat (fire) then crushed with a knife. The juice, obtained by squeezing, is drunk at the rate of one glass per day per adult, until the symptom clears (Efate, 14th March, 1979). In the second case, a section of the external part of the trunk, either fresh or dried, the latter form usually kept in the hut for such events, is boiled in water until obtaining the colouration of tea. (In Vanuatu, the comparison is usually made to tea of Sri Lanka (Ceylon Tea), which is very red). The preparation is then conserved and for effect it is necessary to drink one or two bottles: 50 to 65ml. (Efate, 10th Nobember, 1981)

On Erromango, an aphrodisiac/medicine is prepared by mixing the ashes of the bark of *Pemphis acidula* (a Lythraceae which grows on windy shores) with the ashes of the bark of *Casuarina equisetifolia,* the filao, (which grows on windy shores but farther back from the shoreline) and crushed coconut endosperm (Ipota, 25th July, 1983.) It is obvious that the ingestion of the mixture is useful in one way at least. The ashes come from two plants containing mineral salt in noticeable quantities. This is an interesting point, as the local food is very bland and rarely spiced. *Pemphis acidula* and *Casuarina equisetifolia* also play a symbolic role. The former grows in rather difficult conditions, on emergent reefs, and filao, the latter, is one of the hardest timbers in Vanuatu, and in French is known as ironwood. The crushed coconut does not appear to have any benefit other than to aid the ingestion of the medicine.

Also on Erromango, a preparation against ringworm is made in a similar way, but for external use only. The bark of *Citrus macroptera, Pterocarpus indicus* and another undetermined species are crushed, extracts obtained and incorporated into a crushed coconut mixture. The ointment is applied to the skin with a chicken feather (Ipota, 25th July, 1983).

Another preparation used on Erromango is a remedy against the sickness SEVE, probably corresponding to Bancroft's filariosis. Once again, crushed or grated coconut is mixed with the juice of the crushed medicinal plant (Ipota, 25th July, 1983).

In conclusion the role of the coconut palm in Vanuatu is essentially to furnish useful materials for everyday life in the villages, to furnish an edible heart and most importantly as raw material in the production of copra and oils. It would appear that any medicinal uses are relatively unimportant; being used as a sweetener in diverse preparations, particularly as a remedy for diuresis. More importantly, its uses in this field are strictly symbolic. From previous studies of this species, it is obvious that there is little chance of the discovery of new medicines. Of further interest would be the continuance of the ethnobotanical studies based upon the ideas and thoughts of the Ni-Vanuatu and other inhabitants of the archipelago.

CARYOTA

The first mention of this genus in Vanuatu appears to be by MacGillivray (1854) "... four palms occur, *Cocos, Caryota, Areca* and *Sagus;* the last three are rare and are only found near habitations . . . " It is possible that this genus was introduced to Anatom, about that time, by a merchant, Mr. Paddon, who had opened a business in 1844 (O'Reilly, 1957) and was interested in the local flora. Also, according to O'Reilly, Mr. Paddon had taken, in 1852 or 1853, an exhibition 'Hebridais', to the Sydney Botanic Gardens, where he displayed plants from Vanuatu.

It is not known if MacGillivray collected samples in Vanuatu, but the genus has been reported on Tanna by Schmid, "... *Caryota* sp; Palm with single trunk, leaves over 3m long (18 pairs of pinnae, 20 pairs of pinnules), abundant in forest on steep ferralitic slopes; north of the island at about 150-250m." (Schmid, Flora of Tanna: Supplement, Dec. 1970)

Also "... *Caryota* sp. Tanna, north, in the forests on the slopes above the river at 100-200m. There is doubt about it being indigenous." (Schmid, Phanerogams of Vanuatu, Jan. 1973)

These indications probably correspond to the specimen Schmid 3630 conserved at NOU, 8th Dec. 1970, from Tanna at 100-200m.

Two other specimens of *Caryota* sp. have been collected on Tanna:
- Phillipe Morat 5863, Feb. 1978, Tanna, white-Grass, conserved at NOU and P.
- Igor de Garine, without number, no date. (*Sagus?),* probably 2nd or 3rd Dec. 1977, Tanna, Loanatom, Imeneka: conserved at Port Vila, but determined by Jean-Marie Veillon at Noumea in 1977 as *Caryota* sp. and then by the author on 12th May, 1984 at Port vila.

One of several species of *Caryota* could have been imported into Vanuatu by Mr. Robert Lambert, nurseryman of Melemat, near Devil's Point, up until 1985, prior to the sale of his business to New Zealanders. Mr. Lambert had travelled many times to Central America, particularly to Panama. He told me that he had brought back palm seeds which included *Caryota.* A.C. Smith confirmed the presence of *Caryota* in Vanuatu in 1979.

Utilization:

The palm-heart is consumed on Tanna (Morat, 5863). De Garine does not elucidate more about his '*Sagus.*' The vernacular name is NIP (T. 171)

According to Mr. Robert Lambert, a *Caryota* sp. in his garden causes a skin allergy and any contact with the fruit should be avoided. Possibly this could only be a personal reaction.

In conclusion, it would be interesting to ascertain which species of *Caryota* are growing in Vanuatu; to compare those from Tanna, Anatom and Efate and to note their uses, whether they be for ornamentals or consumption.

ARECA

Areca catechu L.

Codrington (1891) reports that the Mbua or Areca nut is sought after by the people of the Solomon Islands living in the Torres Group for use as a hunger suppressant. There appears to be no narcotic effect on the people who regularly use Mbua which can explain the immunity which they have developed. The active principles in Mbua being alcaloides from tetrahydonicitinic acid; the principle one being arecoline (Paris 1967).

The use of Mbua as a masticatory is unusual in Vanuatu, the only group of people regularly using it are those from the Solomon Islands. I observed its use only once in Port Vila (28th Sept. 1980). Its use appears restricted throughout the archipelago, although Madame D. Bourret noted that (1978) on Motalava the seeds of *Areca catechu* were used as masticatory and worming tablets. These two indications are perfectly in accordance with accepted pharmacology.

The distribution of the genus *Areca* in Vanuatu is unknown as is the origin of the seeds used as masticatory (chewing). This is an ethnobotanical problem and it appears that the habit of betel and of kava are mutually excluded; we could not find any kava cultivated in the Solomons nor any *Areca catechu* being cultivated in Vanuatu.

ROYSTONEA

Roystonea oleracea (Jacq.) O.F. Cook

The presence of *(Areca oleracea* Forster non L.) *Roystonea oleracea* was reported in Vanuatu, on Tanna and Malekula (1774) by Forster (1786). This indication was reported again by Seemann (1868) and Guillaumin (1927) who states in his compendium of 1948 that "The presence of *Areca oleracea* was reported in Vanuatu without precision by Forster; it is evidently not that species of palm" (1948).

After his voyage, MacGillivray reported the presence of *Cocos, Caryota, Areca* and *Sagus (Metroxylon)* on Anatom (Nov. 1853), the last three genera being rare and only near habitations (MacGillivray 1854). The same species of *Areca* is also reported on the island of Futuna by the same author (23rd Dec. 1953).

LIVISTONA

Only one reference to *Livistona* is known, it being from Guilfoyle (1869), where he notes that on the island of Anatom it is abundant in the interior.

CHAMBEYRONIA

Chambeyronia sp. aff. ***macrocarpa*** Vieill.

One species of this genus is reported to have been collected from Vanua Lava by S.F. Kajewski (428, 6th July, 1928) . . . a palm up to 15m tall. According to Guillaumin, it appears to be *Chambeyronia* sp. aff. *macrocarpa* Vieill. (Guillaumin, 1932), with an illustration of the fruit cut lengthways, p. 112.

This description is found again in two other articles by Guillaumin, 1948 and 1954. The fruit is edible according to Guillaumin (1932 and 1954)

CYPHOPHOENIX

Two references indicate the presence of *Cyphophoenix* in Vanuatu. From Guillaumin (1932 and 1948), the sample Kajewski 465 collected on Vanua Lava, could be of this genus, whilst samples collected on Espiritu Santo by Ina Baker (318 and 361) appear similar to the genus. A revision is necessary.

No traditional uses have been recorded for the specimens.

BASSELINIA

One species of *Basselinia* is noted as occurring on Tanna by the ethno-nutritionist, Igor de Garine, whilst passing through Loanatom on December 1977. According to this researcher, his assessment places a species known as NAWUBUK (T. 171) to be a species of *Basselinia.* Its uses are not recorded and there is mention of a lost sample of the species.

UNDETERMINED PALM SPECIES

A number of samples remained undetermined at the time of the compilation of this article. Furthermore, the ethnobotanical studies indicate to us that there may be other species, which are indicated by vernacular names. but grouped together in the bichlamar language.

For example, on Erromango, there occurs the NIVAL which has been placed in *Veitchia* and the NOKIMPAT which is abundant in the Rantop Peninsula at 837m alt in the eastern part of the island. The latter palm has undivided leaves (i.e. not split into leaflets) but fruits which resemble the former (*Veitchia* P. Cabalion 2264). The apical bud of both species is consumed. Other palms with vernacular names include the NOKI (most likely *Cocos nucifera),Metroxylon* which has several names, the two species of *Calamus* and finally the YARPAT which should also be attempted to be collected. There is evidence from the ethnobotanical studies which indicates that separate species may be represented by the NOKIMPAT and the YARPAT.

Concerning *Metroxylon,* there is ambiguity on the islands of Erromango and Malekula where we found the NATANGURA being referred to as NUWAURIET (P. Cabalion, 16th November, 1982). It is distinguished from another species, the NINDRU AMBIH (P. Cabalion 1844) which has a tall trunk, longer leaves and which takes a longer time to bear fruit. Could it be *M. warburgii* or *M. salomonense?* The precise distribution of these two species should be established to allow a better understanding of the relationship between them.

An undetermined palm collected on Vanua Lava produces new leaves which are edible (J.-M. Veillon, 5567). It is known as N'GUP OU HUUP (language similar to SASAR, T 8). It is possibly *Pelagodoxa henryana.*

Another, also collected by J.-M. Veillon (5512), is used in the fabrication of poison arrows. A point of human bone or fish bone is impregnated with poison derived from a decaying carcass. This is then affixed to a stem from the particular palm and this is then extended with a thatch from a species of Poaceae. As the point of the arrow penetrates the victim, the stem of the palm is broken because of the weight of the thatch and the violent reaction of the victim and the point remains intact and not able to be extracted.

In conclusion, it must be noted that a complete systematic study of the palms of Vanuatu has yet to be undertaken, and further study of their uses needs to be initiated to establish efficient methods of conservation and protection.

REFERENCES FOR THE INTRODUCTION TO VANUATU PALMS

(1) GUILLAUMIN A. (1927) Contribution a la flore des Nouvelles-Hebrides. II. Liste des plantes connues. Bull. Soc. bot. Fr., Paris, 74,5 ser., III: 693-712.

(2) GUILLAUMIN A. (1948) Compendium de la flore phanerogamique des Nouvelles-Hebrides. Ann. Mus.Col. Marseille, 55 et 56 annees, 6 ser., 5 et 6 vol.: 5-56.

(3) LEE K.E. (1974) Royal Society and Percy Slades Expedition to the New Hebrides, 1971, Collection Data. Royal Society, London.

(4) OFFICE de la RECHERCHE SCIENTIFIQUE et TECHNIQUE OUTRE-MER. (Port Vila Centre: BP 76, Port Vila, Vanuatu) (Noumea Centre: BP A5, Noumea, New Caledonia)

(5) Directed by Pr H.A. MILLER – Univ. of Central Florida, Orlando.

(6) CABALION P., SAM Ch. et SEOULE S. (1983) Herbiers de Vanuatu representes a Port Vila (PVNH) et a Noumea (NOU) – (comptage) NAIKA, J. Soc. Sc. Nat. Vanuatu, 12: 12-15.

(7) Pers. com. from Dr. HELEN I. ASTON, Senior Botanist at Royal Botanic Gardens and National Herbarium, South Yarra, Australia.

(8) Pers. com. from Pr H.A. MILLER.

REFERENCES FOR VANUATU PALM GENERA

BAKER J.R. (1935). Espiritu Santo, New Hebrides. Geographical J., 85, 3: 209-233 & 230-233

BARRAU J. (1956). L'agriculture vivriere indigene aux Nouvelles Hebrides. J. Soc. Oceanistes, T.xii, numero special consacre aux Nouvelles Hebrides: 181-215

BOURDY G. (1989). Ethnopharmacologie et vie sexuelle a Vanuatu. 2 Vol. These de Sci. Pharm. Montpellier, 27 Avril 1989.

BOURRET D. (1978). Communication personnelle.

BURRET M. (1932). Die Palmengattung Kajewskia Guillaumin und Carpoxylon H. Wendl. et Drude von der Insel Aneityum, Neu-Hebriden. Notizbl der konigl. botan. Gartens und Museums zu Berlin, 11: 578

BURRET M. et POTZIAL E. (1956). Systematische ubersicht uber die Palem, Paper N† XXI, reprinted from Willdenowia, Berlin Dahlem, Brothers.

CABALION P. et MORAT Ph. (1983). Introduction a la vegetation, a la flore et aux noms vernaculaires de l'ile de Pentecote (Vanuatu). J.A.T.B.A., XXX, 3-4; 197-248

CODRINGTON R.H. (1891). The Melanesians, studies in their anthropology and Folklore. 419 p., Oxford Clarendon Press.

ESSIG F.B. (1982). A synopsis of the genus Gulubia. Principes, 26 4: 159-173

FORSTER G. (1786). Florulae insularum australium Prodromus.

GOWERS Sh. (1976). Some common trees of the New Hebrides and their vernacular names. 159 p. Forest. Sect., Dept. of Agric., Port Vila.
GUIART J. (1956). Un siecle et demi de contacts culturels a Tanna, Nouvelles Hebrides, Soc. des Oceanistes. publ. n† 5.
GUILFOYLE W.R. (1869). A botanical tour among the south sea islands. The J. of Botany, London, 7: 117-136
GUILLAUMIN A. (1919). Contribution a la Flore des Nouvelles Hebrides. I. Premisses a la Flore d'Efate. (Recoltes de M. Levat).Bull. Soc. Fr., Paris, 66: 267-277.
GUILLAUMIN A. (1727) Contribution a la Flore des Nouvelles Hebrides II. Liste des plantes connues. Bull. Soc. Bot. Fr., Paris, 74 5 ser., T.III: 693-712.
GUILLAUMIN A. (1932). Contribution to the Flora of the New Hebrides. Plants collected by S.F. Kajewski in 1928 and 1929. J. Arnold Arb., 13: 1-29 & 81-126.
GUILLAUMIN A. (1938). A florula of Espiritu Santo, one of the New Hebrides. J. Linnean Soc. London, Botany, 51, 340: 547-566
GUILLAUMIN A. (1948). Compendium de la Flore Phanerogamique des Nouvelles Hebrides, Ann. Mus. Col. Marseille, 5 & 6 Vol. : 5-56
GUILLAUMIN A. (1954) Les plantes utiles des Nouvelles Hebrides. J.A.T.B.A., I, 7-9: 293-297, & 10-12: 453-460
JARDIN Cl.(1974). Kulu, Kuru, Uru. Lexique des noms de plantes alimentaires dans le Pacifique Sud. Cahiers d'informat. CPS n† 35, 231p.
KAJEWSKI S.F. (1930). A plant collector's notes on the New Hebrides and Santa Cruz Islands. J. Arnold Arb., 11: 172-180
MacCURRACH J.C. (1960). Palms of the World. 290 p. Harpers & New York, re-edition 1976, Horticultural Books, Stuart, Florida.
MacGILLIVRAY J. (1854) (Lettre de-, naturaliste du Herald, commande par le Capitaine Denham, datee de Sydney, 3 mars 1854) in: Hook. J. Bot., 6: 353-363
MOORE H.E. et FOSBERG (1956). Gentes Herbarum, Vol. 8, 6: 458-466
O'REILLY P. (1957) Hebridais. Repertoire bio-bibliographique des Nouvelles Hebrides. 290 p., Publ. Soc. Oceanistes,6
O'REILLY P. (1958). Bibliographie des Nouvelles Hebrides, 304 p., Publ. Soc. Oceanistes. 8
PARIS R.R. et MOYSE H. (1967). Precis de Matiere Medicale. T. II, 511 p., Masson, Paris. Pichi Sermolli
SARLIN P. (1950). La foret et le sol du nord de Santo (Nouvelles Hebrides), Bois et Forets des Tropiques, 13: 4-13
SCHMID M. (1965). Especes Vegetales observees a Vate. 26 p., ORSTOM Noumea.
SCHMID M. (1970). Florule d'Anatom, 53 p., ORSTOM Noumea
SCHMID M. (1971) Florule d'Anatom, Supplement, 21 p. ORSTOM Noumea
SCHMID M. (1971) Florule d'Anatom, 2 Supplement, 13 p. ORSTOM Noumea
SCHMID M. (1970). Flo:ul de Tanna, 40 p. ORSTOM Noumea.
SCHMID M. (1970) Florule de Tanna, Supplement I, 21 p., ORSTOM Noumea.
SCHMID M. (1973). Florule d'Erromango, 52 p. ORSTOM Noumea
SCHMID M. (1973). Phanerogames des Nouvelles Hebrides , 54 p. ORSTOM Noumea.
SCHMID M. (1973). Especes de Vegetaux superieurs observes a vate, Nouvelles Hebrides, 42 p. ORSTOM Noumea.
SCHMID M. (1974). Florule de Pentecote, 25 p. ORSTOM Noumea.
SEEMANN B. (1868-1873). Flora Vitiensis. 453 p., Londres; (Reed, Cramer, Vaduz, 1977)
SMITH A.C. (1979). Flora Vitiensis Nova, Vol. 1, 495 P.
STEENIS C.G.G.J. (van-) et BALGOOY M.M.J. (van-), (1976). Pacific Plant Areas, Vol. 2, Blumea, Vol. V, 312 p.
TRYON D.T. (1976). New Hebrides Languages: An internal Classification. Pacific Linguistics Series C. n† 50, ANU, Canberra.
WALTER A. et BOURDY G. (1987). Fibres et Liens. Quelques plantes de Vanuatu utilisees en technologie traditionelle. Naika, J. Vanuatu Natural Science Society, 26: 9-12
WALTER A. et CABALION P. A paraitre –
WEIGHTMAN B. (?). Ouvrage de synstese sur l'agriculture a Vanuatu.
WILLIS J.C. (1897, rev. 1985). A dictionary of the flowering Plants & Ferns, 1245 p., Cambridge University Press.

NEW CALEDONIA PALMS IN CULTIVATION

The cultivation of New Caledonia palms has proven to be a challenge for most growers and apart from *Chambeyronia macrocarpa* and to a lesser extent *Kentiopsis oliviformis,* there are very few successfully in cultivation. The following notes and comments may introduce and hopefully solve some of the problems associated with cultivating this difficult group.

Gerald McKiness of Georgia, U.S.A. writes -

"I must confess I have had very poor success at growing the New Caledonia palms. I did have success with a few though. I was able to get some fresh seed of *Basselinia gracilis* and was able to keep them thriving for six years. The plants were very healthy and were in a well drained soil mix containing pine mulch (one- third), perlite (one-third) and sandy loam (one-third).

"Another palm I had success with was *Burretiokentia vieillardii.* I grew some to a very nice three gallon container size. However, I did lose a fair percentage of the seedlings during the winters. It is my belief that the deaths were due to rot caused from cool damp conditions where the excess moisture could not be evaporated. I believe also that if these palms were grown in a similar way as the culture recommended for *Phalaenopsis* and *Odontoglossum* species of orchids that much success would be realised. Get the seedlings past the first three years like this and then bring them out. A well drained, organically rich soil is very important. I was able to germinate *Cyphosperma balansae, Basselinia pancheri, Cyphokentia macrostachya, Brongniartikentia vaginata, Actinokentia divaricata* and others using this mix. I was able to get most seedlings to survive to the first winter before I lost them.

Most of my plants would be grown outdoors in south Florida under tree canopy or under saron shadecloth. The forest palms which require a more acid soil would be treated with manure and iron chelate at least once or twice a year. The death of the seedlings as I see it was not in the soil condition or amount of irrigation. I really believe it is the amount of warm mist needed around the plant as might be produced in a mist propagation house. I would also emphasise regular treatments of a systemic fungicide and some air movement. Ideal conditions to be found in a good orchid house."

From Cairns, Dusan Balint writes:

"Due to the difficulties in the past of obtaining fresh, viable seed from New Caledonia, I have not grown many of the very interesting palms. I list below those which I am growing at present.

Basselinia gracilis - is a delightful little palm presenting no difficulties to the grower provided that it is grown in a very well drained medium. I found it to be very slow growing initially.

Brongniartikentia vaginata - equally slow growing and susceptible to damping off.

Chambeyronia macrocarpa - this palm seems to tolerate a large geographical distribution growing well in subtropical Sydney or tropical highlands and here in tropical Cairns.

Chambeyronia lepidota - similar to *C. macrocarpa.*

Burretiokentia vieillardii - a few young plants of this palm I obtained three years ago, which had grown to a six inch pot size, rotted off last summer.

I am under the impression that all New Caledonia palms would prefer a climate slightly cooler and less humid than we have here in Cairns."

Scott A. Lucas, Curator of Living Collections at the Pacific Tropical Botanical Garden in Hawaii writes:

"There are not many (New Caledonia) species in cultivation here. I am sure that the reason does have to do with the major soil differences between ours which is acid and volcanic in origin, and New Caledonia's which is alkaline and serpentine in origin. Nevertheless, we have succeeded in establishing a few species in cultivation here. These include the following:

Basselinia pancheri
Brongniartikentia lanuginosa
Burretiokentia vieillardii
Kentiopsis oliviformis
Moratia cerifera

"The most critical time seems to be the few months immediately following germination. The list above represents a very small percentage of the fifty or so plant species from New Caledonia which we have attempted growing over the years. We have never really made an attempt to amend our soils for the planting of New Caledonia plants and we seem to have equal trouble getting plants other than palms into cultivation here as well. We have made a number of unsuccessful attempts at getting *Pritchardiopsis* to grow here and will no doubt keep trying."

Pauline Sullivan, of the International Palm Society lists the following as growing in her gardens in Ventura, California:

Chambeyronia macrocarpa	*Basselinia gracilis*
Burretiokentia vieillardii	*Actinokentia divaricata*
Kentiopsis oliviformis	*Veillonia alba*

Pauline surveyed some fellow palm growers within her area (Southern California) and supplied the following:

The species cultivated in Southern California include, as well as those above, *Cyphosperma balansae* and *Burretiokentia hapala.* It appears that *Chambeyronia macrocarpa* is very adaptable to the Southern California climate.

Pauline adds, quote:- "This last winter was unusually cold. Twice the temperature dropped to freezing with many nights around 2-5°C. None of my New Caledonia palms showed any cold damage. All my New Caledonia palms I give chelated iron, an abundance of water and fertilize regularly. My soil is clay."

From the Solomon Islands writes Geoff Dennis, quote:

"I am successfully growing *Burretiokentia vieillardii, Chambeyronia macrocarpa* and *Actinokentia divaricata* in our local (Honiara) calcareous soils at about 170 ft. elevation a.s.l. from seeds originally collected by Don Hodel on behalf of the International Palm Society in the mid 1970's. All species are just commencing to develop trunks.

"This is very satisfactory in view of our ever warm climatic conditions compared with New Caledonia's sub-tropical climate with fairly low night and mid-year temperatures and the fact that the first and last named species are normally found only on soils overlying schists and serpentine rocks. *Chambeyronia macrocarpa* is widely distributed throughout its home island and is not so specific in its growing requirements, doing well also in Florida, U.S.A., which also has mainly calcareous soil conditions in the warmer southern growing region. Other seedlings raised from Don Hodel's gatherings, including various species of *Basselinia, Mackeea magnifica,* etc. failed to survive although I believe that some may thrive if provided with suitable growing media. Certainly there are all kinds of soil types accessible in our neighbouring bush areas inland, including ultra- basics, and I am in fact raising a few recently acquired seedlings of *Burretiokentia vieillardii, Chambeyronia macrocarpa, Cyphokentia macrostachya* and *Cyphophoenix elegans* with success to date."

Ken Foster, Past President of the International Palm Society writes -

"My first palm collecting trip to New Caledonia was in 1977. I again collected in 1979. I list below 15 species of palms that I personally collected on these two trips and the results:

Chambeyronia macrocarpa - easy to germinate and easy to grow.
Kentiopsis oliviformis - easy to germinate and easy to grow.
Burretiokentia vieillardii - easy to germinate and easy to grow.
B. hapala - easy to germinate and easy to grow.
Brongniartikentia vaginata - seeds germinated poorly, but plants healthy, but very slow.
Veillonia alba - easy to germinate but slow growing.
Clinosperma bracteale - seeds hard to germinate and very slow to grow.
Mackeea magnifica - seeds easy to germinate but very slow growing, plants robust.
Basselinia gracilis - seeds very easy to germinate, plants slow.
Cyphophoenix elegans - seeds hard to germinate, plants slow but robust.
C. nucele - seeds easy to germinate and easy to grow, very robust.
Cyphosperma balansae - seeds hard to germinate and plants slow.
Campecarpus fulcitus - seeds did not germinate, were not quite fully ripe.
Cyphokentia macrostachya - seeds did not germinate, were fully ripe.
Actinokentia divaricata - seeds easy to germinate, but plants died after several years, were two feet tall.

REFERENCES FOR PALMS OF THE SOUTH-WEST PACIFIC

BARRAU, J. (1960). Principes Vol.4. The Sago Palms.

BLOMBERY, A. & RODD, T. (1982). Palms. Angus & Robertson.

CABALION, P. & MORAT, P. (1983). Journ. D'Agric. Trad. et de Bota. Appl. 30-(34). Introduction a la Vegetation, a la Flore et aux noms Vernaculaires de L'Ile de Pentecote (Vanuatu).

CABALION, P. (1984) Journal de la Societe des Oceanistes. Les Noms des Plantes Bichlamar Origines, Formations et Determinations Botaniques.

COLEMAN, P.J. Edited by (1973). The Western Pacific. University of Western Australia Press.

DANIOU, C. & ELLIS, C. (1984). Vanuatu. Vanuatu National Institute of Technology.

DRANSFIELD, J. (1979). Malayan Forest Records No.29. A Manual of the Rattans of the Malay Peninsula. Forest Dept. Ministry of Primary Industries, Malaysia.

DOWE, J.L. (1988). Palms & Cycads No.18. The Rediscovery of *Carpoxylon macrospermum.*

DOWE, J.L. and UHL, N.W. (1989) Principes Vol. 33, Carpoxylon macrospermum

ESLER, A.E. (1969). Bulletin of the Wellington Botanical Society 36: 18-23. Leaf Fall and Flowering of Nikau.
(1974). New Zealand's Nature Heritage, Part 19. The Nikau Palm. Paul Hamlyn Ltd.

ESSIG, F.B. (1982). Principes Vol.26. A Synopsis of the Genus *Gulubia.*

FLEMING, C.A. (1979). The Geological History of New Zealand and Its Life. Auckland University Press.

HUTTON, I. Discovering Australia's World Heritage. Lord Howe Island. Conservation Press.

JAFFRE, T. & VEILLON, J.-M. (1989). Morphologie, Distribution et Ecologie des Palmiers de Nouvelle-Caledonie.

JOHNS, R.J. & HAY, A.J. (1984). A Guide to the Monocotyledons of Papua New Guinea Part 3. Office of Forests.

JONES, D. (1984). Palms in Australia. Reed Books.

LANGLOIS, A.C. (1976). Supplement to Palms of the World. University of Florida.

LOUKAKIS, A. (1984). Norfolk, An Island and Its People. Rigby.

McCURRACH, J.C. (1960). Palms of the World. Harper & Bros., New York.

MOORE Jr., H.E. (1957). Gentes Herbarum Vol.8. Fasc.7. Art.21. *Veitchia* Art.22 *Neoveitchia.*
(1958). Principes Vol.2. The Cultivated Veitchias.
(1966). Principes Vol.10. Palm Hunting Around the World VI. New Caledonia and Fiji.
(1969). Principes Vol.13. A Synopsis of the Genus *Physokentia.*
(1969). Principes Vol.13. New Palms from the Pacific.

MOORE Jr., H.E., PHILLIPS, R.H. & VODONAIVALU, S. (1982). Principes Vol.20. Additions to the Palms of Fiji.

MOORE Jr., H.E. & UHL, N.W. (1982). The Botanical Review. Vol.48 No.1. Major Trends of Evolution in Palms. New York Botanical Garden.
(1984). Allertonia Vol.3 No.5. The Indigenous Palms of New Caledonia. Pacific Tropical Botanical Garden.

MOORE, L.B. & EDGAR, E. (1970). Flora of New Zealand Vol.2. Government Printer, Wellington.

NORFOLK ISLAND FLORA & FAUNA SOCIETY (1987). Monthly Nature Notes.

PARHAM, M.A. Fijian Plants. Society of Science & Industry.

PICKARD, J. (1983). Cunninghamia Vol.1(2). Vegetation of Lord Howe Island. Royal Botanic Gardens, Sydney.

PLAN OF MANAGEMENT OF THE NORFOLK ISLAND NATIONAL PARK (1986).

RAUWERDINK, J.B. (1986). Principes Vol.30. An Essay on *Metroxylon,* The Sago Palm.

SCHMID, M. (1973/74). Nature Caledonie. Palmiers.

SCHMID, M. (1981). Fleures et Plantes de Nouvelle-Caledonie. ORSTOM. Les Editions du Pacifique.

SMITH, A.C. (MOORE Jr., H.E.) (1979). Flora Vitiensis Nova. Family 39. Arecaceae. Pacific Tropical Botanical Garden.

SMITH, J.M.B. Edited by (1982). A History of Australasian Vegetation, McGraw-Hill, Sydney.

STANLEY, D. (1986). South Pacific Handbook. 3rd Edition. Moon Publishers.

SYKES, W.R. (1977). New Zealand DSIR. Bulletin 219. Kermadec Islands Flora. DSIR Wellington.

TUCKER, R. (1988). Palms of Subequatorial Queensland. Palm & Cycad Societies of Australia, Publication Fund.

UHL, N.W. & DRANSFIELD, J. (1987). Genera Palmarum. The L.H. Bailey Hortorium and the International Palm Society.

For Churchill (1973), Heckel (1972), Lange (1982), Partridge (1976), and Singh (1982), see Smith J.M.B., edited by (1982).

For Oliver (1911), see Sykes (1977).

SOURCES OF ILLUSTRATED MATERIAL

Dried specimens from the ORSTOM Herbarium (NOU) Noumea, New Caledonia:

Actinokentia divaricata Hoff 425
Alloschmidia glabrata (?)
Basselinia deplanchei Stone 14964
B. iterata Moore 10561
B. sordida MacKee 17367
B. tomentosa Veillon 3928
B. porphyrea Morat 5974
B. velutina Moore 9964
B. vestita McPherson 3033
Brongniartikentia lanuginosa Veillon 3563
Burretiokentia hapala MacKee 36036
B. vieillardii MacKee 78505
Campecarpus fulcitus Moore 9344
Chambeyronia lepidota Schmid 7617
Clinosperma bracteale MacKee 20648
Cyphokentia macrostachya MacKee 31836
Cyphophoenix elegans Schmid 7609
C. nucele Veillon 4172
Cyphosperma balansae Moore 10073
Kentiopsis oliviformis Schmid 1222
Lavoixia macrocarpa (?)
Mackeea magnifica Moore 10054
Moratia cerifera (?)
Veillonia alba Schmid 1615

Dried specimens from Tagabe Herbarium (PVV), Port Vila, Vanuatu:

Calamus sp. aff. *vitiensis* Moore 1976
Clinostigma harlandii Moore

Fresh specimens from Dowe 1987/88

Calamus sp. aff. *vitiensis*
Gulubia cylindrocarpa
Howea belmoreana
H. forsteriana
Metroxylon warburgii
Veitchia macdanielsii
V. montgomeryana
V. winin

Fresh specimens from other sources:

Balaka longirostris Phillips, Fiji
Clinostigma exorrhizum Phillips, Fiji
Physokentia rosea Phillips, Fiji
Rhopalostylis sapida Endt, New Zealand
Balaka seemannii I.P.S. Seedbank, U.S.A.
Hedyscepe canterburyana I.P.S. Seedbank, U.S.A.
Lepidorrachis mooreana I.P.S. Seedbank, U.S.A.
Veitchia sessilifolia I.P.S. Seedbank, U.S.A.

PHOTOGRAPH CREDITS

All unacknowledged photographs by the author; Otherwise as follows:

(AD) Andre' de Does, Brisbane
(BDS.NZGS) B. D. Scott, New Zealand Geological Survey
(CT) Chris Tiffin, Brisbane
(DE) Dick Endt, New Zealand
(GC) Greg Cuffe, Brisbane
(GH) George Harriman, Brisbane
(HT) Helen Tiffin, Brisbane
(JP) Jeanne Price, Brisbane
(JS) J. Sorenson, New Zealand
(LHBH) L.H. Bailey Hortorium, Cornell University, U.S.A.
(O) ORSTOM, Noumea, New Caledonia
(PC) Pierre Cabalion, Strasbourg, France
(RF) Rosebud Farm, Kuranda, Queensland
(SG) Shelley Gage, Brisbane

INDEX OF GENERA AND SPECIES

THE SOCIETIES

THE PALM & CYCAD SOCIETIES OF AUSTRALIA is the title given to the body which includes the affiliated societies listed below: –

PALM & CYCAD SOCIETY OF AUSTRALIA – The Southern Queensland Group based in Brisbane.

NORTH QUEENSLAND PALM SOCIETY – based in Townsville.

PALM SOCIETY OF THE NORTHERN TERRITORY – based in Darwin.

SUNSHINE COAST PALM SOCIETY – based in Nambour.

PALM & CYCAD SOCIETY OF MACKAY – based in Mackay.

PALM SOCIETY OF NEW SOUTH WALES – based in Sydney.

PALM & CYCAD SOCIETY OF SOUTH AUSTRALIA – based in Adelaide.

PALM & CYCAD SOCIETY OF WESTERN AUSTRALIA – based in Perth.

The objects of the Societies include the promotion of the culture and propagation of palms and cycads and the promotion and dissemination of scientific and cultural knowledge.

In December 1989 membership stood at close to 1300 which included 200 overseas members from 10 countries with a predominance resident of the U.S.A.

The Societies have expanded through the continued affiliation of associated societies since 1983, and this affiliation is reflected in the Societies' name, representing of grouping of semi-autonomous bodies. P.A.C.S.O.A. has also been affiliated with the International Palm Society based in the U.S.A., since 1984.

Benefits of membership include: –

- PALMS & CYCADS the Societies' quarterly colour magazine
- Branch group newsletters and participation in branch group activities
- Participation in the Societies' SEEDBANK which in 1989 distributed over 15,000 packets of seed of almost 150 species of palms and cycads
- Participation in the Societies' Cycad Pollen Bank
- Discounts on palm and cycad publications through the Societies' BOOKSTORE

This publication was funded by members' donations and fundraising activities.

Write for a membership application form and further details of the Societies to the Secretary at:

PALM & CYCAD SOCIETIES OF AUSTRALIA
P.O. BOX 1134
MILTON QLD. 4064
AUSTRALIA